AF720126

Urban Engineering

Urban Engineering

Edited by
Alex Austin

STATES
ACADEMIC PRESS
www.statesacademicpress.com

Published by States Academic Press,
109 South 5th Street,
Brooklyn, NY 11249, USA

Copyright © 2022 States Academic Press

This book contains information obtained from authentic and highly regarded sources. Copyright for all individual chapters remain with the respective authors as indicated. All chapters are published with permission under the Creative Commons Attribution License or equivalent. A wide variety of references are listed. Permission and sources are indicated; for detailed attributions, please refer to the permissions page and list of contributors. Reasonable efforts have been made to publish reliable data and information, but the authors, editors and publisher cannot assume any responsibility for the validity of all materials or the consequences of their use.

Trademark Notice: Registered trademark of products or corporate names are used only for explanation and identification without intent to infringe.

ISBN: 978-1-63989-550-2

Cataloging-in-Publication Data

Urban engineering / edited by Alex Austin.
p. cm.
Includes bibliographical references and index.
ISBN 978-1-63989-550-2
1. Municipal engineering. 2. Urbanization. 3. City planning. 4. Public works.
5. Engineering. I. Austin, Alex.

TD159 .U73 2022
628--dc23

For information on all States Academic Press publications
visit our website at www.statesacademicpress.com

STATES
ACADEMIC PRESS

Contents

Preface

I am honored to present to you this unique book which encompasses the most up-to-date data in the field. I was extremely pleased to get this opportunity of editing the work of experts from across the globe. I have also written papers in this field and researched the various aspects revolving around the progress of the discipline. I have tried to unify my knowledge along with that of stalwarts from every corner of the world, to produce a text which not only benefits the readers but also facilitates the growth of the field.

Urban engineering, also known as municipal engineering, is a branch of engineering concerned with urban infrastructure. It is a sub-discipline of civil engineering and specializes in residential and commercial development, planning & design, marine construction design and public works design. It makes use of science, art and engineering to design, construct and maintain an urban environment. This includes streets, sidewalks, water supply networks, sewers, street lighting, waste management, storage depots, public parks, etc. It may also include electrical, telecommunications, garbage collection and bus service networks. Some of these disciplines overlap with other civil engineering domains. Urban engineering focuses on the coordination of these infrastructure networks and services, as they are often built together and managed by the same authority. This book traces the progress of this field and highlights some of its key concepts and applications. The topics covered herein deal with the core subjects of urban engineering. For someone with an interest and eye for detail, this book covers the most significant topics in this field.

Finally, I would like to thank all the contributing authors for their valuable time and contributions. This book would not have been possible without their efforts. I would also like to thank my friends and family for their constant support.

Editor

1

Research on Urban Engineering Applying Location Models

Carlos Alberto N. Cosenza, Fernando Rodrigues Lima, César das Neves
Federal University of Rio de Janeiro (UFRJ)
cosenza@pep.ufrj.br, frlima@poli.ufrj.br, cdn@poli.ufrj.br
Brazil

1. Introduction

This chapter presents a methodology for spatial location employing offer and demand comparison, appropriate for urban engineering research. The methods and techniques apply geoprocessing resources as structured data query and dynamic visualization.

The theoretical concept is based on an industrial location model (Cosenza, 1981), which compares both offer and demand for a list of selected location factors. Offer is detected on location sites by intensity levels, and demand is defined from projects by requirement levels. The scale level of these factors is measured by linguistic variables, and operated as fuzzy sets, so that a hierarchical array of locations vs. projects can be obtained as result. The array is normalized at value = 1 to indicate when demand matches offer, which means the location is recommended.

The case study is solved with geoprocessing tools (Harlow, 2005), used to generate data for a mathematical model. Spatial information are georeferenced from data feature classes of cartographic elements on city representation, as administration boundaries, transportation infrastructure, environmental constrains, etc. All data are organized on personal geodatabase, in order to generate digitalized maps associated to classified relational data, and organized by thematic layers. Fuzzy logic is applied to offer and demand levels, translating subjective observation into linguistic variables, aided by methods for classifying quantitative and qualitative data in operational graduations. Fuzzy sets make level measuring more productive and contributes for a new approach to city monitoring methods. Our proposition is to apply this model to urban engineering, analysing placement of projects that impact on urban growth and development. To operate the model, we propose to use as location factors the environmental characteristics of cities (generic infra-structure, social aspects, economical activities, land use, population, etc).

2. A Location Model

Location models have been used to study the feasibility of projects in a large range of possible sites, and can be applied in macro and micro scale. Macro scale location deals with general and specifics factors, in order to show hierarchical ranking of possibilities. Micro scale location studies come in sequence to choose the most suitable place of a macro studied output, based on local characteristics of terrain, facilities, transportation, population, general services and environmental constrains.

One approach for location problem solving is based on cross analysis (ex: offer vs. demand) of general and specific factors. General factors are important for most projects, and their lack is not imperative for excluding a location site. These factors are related to infrastructure or to some support element that is part of external economies.

Specific factors are essential for some kind of projects, and their absence or deficiency on requested level invalidates the location site. These factors are often related to natural resources, climate, market, etc.

As general and specific factors are not immutable along time, future changes, such as strategic interventions or incoming projects, must also be considered and inputted as part of offer measurement.

The macro location studies here presented are based on offer vs. demand factors and first took place in Italy, with SOMEA research (Attanasio, 1974) to improve balanced development of south and north Italian regions. Their model used a crispy math formulation for the offer vs. demand comparison (Attanasio, 1976), and latter researchers of COPPE/UFRJ (Cosenza, 1981) built a fuzzy approach for this question.

Recently, fuzzy math was applied to find locations for Biodiesel fuel industrial plants and related activities, such as planting and crushing (Lima et al., 2006). The study was territorially segmented in municipalities, so offer level of location factors was measured for each city of Brazil. The government plan for Biodiesel is directed to join economics and social benefits to low-income population, so location studies in this case must deal with a large set of factors, such as agricultural production, logistics and social aspects.

Therefore, the analysis of the multiple facets involved in this kind of study is quite complex. In this sense, the used methodology tries firstly to identify locations potentialities for subsequent evaluation. In the last stage, not only the location options should be considered, but also the project scale and the costs of logistics.

It should be also observed that any methodological proposition cannot be dissociated from the availability and quality of the data for its full application. This means that the propositions of any project can suffer possible alterations along the time, so other aspects not predicted in the model should be analyzed according to the available secondary data.

3. The Mathematical Model

The concept of Asymmetric Distance (AD) does not satisfy the restrictions of Euclydean Algebra and cannot capture the further richness that makes possible to establish a more strict hierarchy. Then, the model was structured in order to evaluate location alternatives using fuzzy logic. The linguistic values are utilized to give rigorous hierarchy by decision-planner under fuzzy environment. In this research a specific fuzzy algorithm was proposed to solve the project site selection.

The first step is facing the demand situations and those of territorial supplying of general factor (basically infra-structure).

Assuming $A = (a_{ij})_{h\times m}$ and $B = (b_{jk})_{n\times m}$ matrices that represent, respectively, the demand of h types of projects relatively to n location factors, and supplying factors represented by m location alternatives.

Assuming $F = \{f_i \mid 1, ..., n\}$ is a finite set of general location factors shown generically as f. Then, the fuzzy set $\tilde{A}$ in f is a set of ordinate pairs:

$$\tilde{A} = \{(f, \mu_{\tilde{A}}(f) \mid f \in \bar{F}\} \quad (1)$$

$\tilde{A}$ is the fuzzy representation of the demand matrix $A = (\mu_{ij})_{h \times m}$ and $\mu_{\tilde{f}}$ is the membership function representing the level of importance of the factors:

Critical - Conditional - Not very conditional – Irrelevant

Likewise, if $\tilde{B} = \{(f, \mu_{\tilde{B}}(f)) \; f \in F\}$ where $\tilde{B}$ is the fuzzy representation of the B supplying matrix and $\mu_{\tilde{B}}(f)$ is the membership function representing the level of the factors offered by the different location alternatives:

Excellent - Good - Fair – Weak

The $\tilde{A}$ matrix is requirement matrix that means that the $\tilde{A}$ set does not have the elements but shows the desired f_i's that belong only to set $\tilde{B}$, defining its outlines, scales levels of quality, availability and supply regularity.

The $\tilde{B}$ matrix with the f_i's satisfies $\tilde{A}$ for proximity. f_1 in the $\tilde{A}$ set is not necessarily equal to f_1 available in $\tilde{B}$. On choosing an alternative, $\tilde{A}$ assumes the values of elements in $\tilde{B}$.
Considering $A = \{a_i / i=1, ..., m\}$ the set of demands in different types of general or common factors for projects (see Table 1), $A_1, A_2, ..., A_m$ are demands subsets and $a_1, a_2, ..., a_m$ different levels of attributes required by the projects.

	f_1	f_2	f_j	f_n
A_1	a_{11}	a_{12}	a_{1j}	a_{1n}
A_2	a_{21}	a_{22}	a_{2j}	a_{2n}
....	....	....	....	....
A_j	a_{j1}	a_{j2}	a_{jj}	a_{jn}
A_m	a_{m1}	a_{m2}	a_{mj}	a_{mn}

Table 1. F_{ij} Factor Demand for Projects

Considering $B = \{b_k \mid k=1, ...,m\}$ the set of location alternatives, where $F = \{f_k \mid k=1, ...,m\}$ is inserted, and represents the set of common factors to several projects (see Table 2), $B_1, B_2, ..., B_m$ is the set of alternatives; $f_1, f_2, ..., f_n$ is the set of factors; $b_1, b_2, ..., b_n$ is the level of factors supplied by location alternatives; and b_{jk} the fuzzy coefficient of the k alternative in relation to factor j.

	Alternatives			
	B_1	B_2	B_j	B_n
f_1	b_{11}	b_{12}	b_{1k}	b_{1m}
f_2	b_{21}	b_{22}	b_{2k}	b_{2m}
....	....	....	....	....
f_j	b_{j1}	b_{j2}	b_{jk}	b_{jm}
f_n	b_{n1}	b_{n2}	b_{nk}	b_{nm}

Table 2. F_{ij} supplying of location alternatives

On trying to solve the problem already figured out on the use of asymmetric distance (AD) and increase the accuracy of the model for the two generic elements a_{ij} and b_{jk}, the product $a_{ij} \otimes b_{jk} = c_{ik}$ is achieved through the operator presented by Table 3, where c_{ik} is the fuzzy coefficient of the k, alternative in relation to an i project, $0^+ = 1/n!$ and $0^{++} = 1/n$ (with n = number of considered attributes) are the limit in quantities and are defined as infinitesimal and small values (>0). Actually, there is an infinite number of values c_{ik} in the interval [0, 1].

	$a_{ij} \otimes b_{jk}$	0	.	.	.	1
Demand for Factors (d)	0	0^+	.	.	.	0^{++}
	.		1			
	.			1		
	.				1	
	1	0	.	.	.	1

Table 3. Supplying Factors (S)

Assuming $a_{ij} = b_{jk}$ the indicator =1, when $b_{jk} > a_{ij}$ the derived coefficient is >1, and when $a_{ij} > b_{jk}$ the fuzzy coefficient is zero (in rigorous matrix) if there is no requirement for a determined factor, but there is a supplying. The fuzzy values are those mentioned above.
In not rigorous matrix $a_{ij} > b_{jk}$ imply in $0 \le c_{ik} < 1$.
Two operators were considered with the same results:
i) not classical fuzzy operation (Table 4);
ii) memberships relation (Table 5).

		supply of factors				
	$a_{ij} \otimes b_{jk}$	0	.	$\mu_{\tilde{B}_i}(x)$	.	1
Demand by Factors	0	0^+	.	.	.	0^{++}
	.		1	$1+\left[\mu\tilde{B}(x)-\tilde{A}(x)\right]$		
	$\mu_{\tilde{A}_i}(x)$			1		
	.	$1+\left[\mu\tilde{B}(x)-\tilde{A}(x)\right]$	1			
	1	0	.	.	.	1

Table 4. Not classical fuzzy

			Weak	Fair	Good	Excellent
		0	$\mu_{B_1}(x)$	$\mu_{B_2}(x)$	$\mu_{B_3}(x)$	$\mu_{B_4}(x)$
	0	1/n!	1/(n-1)	1/(n-2)	1/(n-3)	1/n
Irrelevant	$\mu_{A_1}(x)$	-0,04	1	$1 + \mu_{B_1}(x)/n$	$1 + \mu_{B_2}(x)/n$	$1 + \mu_{B_3}(x)/n$
Not very conditional	$\mu_{A_2}(x)$	-0,16	$\frac{\mu_{B_1}(x)}{\mu_{A_2}(x)}$	1	$1 + \mu_{B_1}(x)/n$	$1 + \mu_{B_2}(x)/n$
Conditional	$\mu_{A_3}(x)$	-0,64	$\frac{\mu_{B_1}(x)}{\mu_{A_3}(x)}$	$\frac{\mu_{B_2}(x)}{\mu_{A_3}(x)}$	1	$1 + \mu_{B_1}(x)/n$
Critical	$\mu_{A_4}(x)$	-1,00	$\frac{\mu_{B_1}(x)}{\mu_{A_4}(x)}$	$\frac{\mu_{B_2}(x)}{\mu_{A_4}(x)}$	$\frac{\mu_{B_3}(x)}{\mu_{A_4}(x)}$	1

Table 5. Memberships relation

Among *n* considered attributes in the several applications, the most frequent ones and those of highest level of support were:

a) elements linked with the cycle of production or service;

b) elements related to transportation and logistics;

c) services of industrial interest;

d) communication;

e) industrial integration;

f) labor availability;

g) electric power (regular supply);

h) water (availability and regular supply);

i) sanitary drainage;

j) general population welfare;

k) climatic conditions and fertility of soil;

l) capacity of settlement ;

m) some other restrictions and facilities related to industrial installation;

n) absence of natural resources that is required by some kind of projects, etc.

The following example of degrees and weights for the i project (Table 6) makes clear the opposition between demand requirements and the conditions of each offering factors.

It can be observed that the operations $O_d \otimes O_s \neq 0$ and $O_D \otimes 1_s \neq 0$ model concerning the hierarchical arrangement of alternatives that do not permit the penalizing of an area that does not have a non-demanded factor or those areas that show more factors than those required, but they can satisfy other requirements and be able to generate external economies.

FACTORS	b_{jk} (Degrees for the k_i alternatives)			a_{ij} (Importance for possibilities)
	B_1	B_2	B_3	
f_1	Weak	Weak	Excellent	Conditional
f_2	Weak	Good	Good	Critical
f_3	Good	Good	Good	Critical
f_4	Weak	Good	Good	Not very conditional
f_5	Fair	Weak	Weak	Irrelevant
f_6	Excellent	Good	Excellent	Conditional
f_7	Good	Excellent	Good	Critical

a_{ij} : fuzzy coefficient of the degree of importance of factor j related to the i project, and
b_{jk} : fuzzy coefficient that results from the level of the factor related to the k area

Table 6. Example of degrees and weights for the i project

Assuming $A^* = (a^*_{ij})_{mxn'}$, the demand matrix of i types of project related to n' specific location factors. Concerning the use of the A matrix, all factors are critical, and for the activities concerning raw materials, these characteristics can be defined by means of the results:

1. Relation product weight / raw material weight
2. Perishable raw materials
3. Relation factor freight / product freight
4. Relation freight factor / factor cost, etc

$\tilde{A}^* = \{f, \mu_{\tilde{A}}{}^*(f) \in F\}$ is the fuzzy representation of the A* matrix.

Assuming $B^* = [bij]_{n'.m}$ the territorial supplying matrix of n' specific location factors of i kind of project, concerning specific resources or any other specific conditioning factor, and $\Gamma = [\gamma_{ik}]_{mxq} = C \oplus C^*$, where the aggregation of values (gamma operation) concerning the activities on specific resources is achieved by Table 7 (with $\tilde{c}_{ik}$ = fuzzy coefficient).

		$\tilde{c}_{ik}$ >0	$\tilde{c}_{ik}$ 0
$\tilde{c}^*_{ik}$	0	0	0
	>0	$c_{ik}+c^*_{ik}$	c^*_{ik}

Table 7. Aggregation operator

The $A = [\lambda_{ij}]_{mxn\Sigma}$ matrix results from that defines the demand profile for the location effect, where: $n_\Sigma = n + n'$.

Assuming □ = $(e_{il})_{h \times h}$ is the diagonal matrix, so that $e_{il} = \begin{cases} 0, & \text{if } i \neq 1 \\ 1/\sum_{j=1}^{n\Sigma} a_{ij}, & \text{if } i = 1 \end{cases}$

$\Delta = [e \times F] = [\delta_{ik}]$ can still be defined as the representative matrix of the location possibilities of the h types of projects in the m alternatives, now represented by indices related to demanded location factors. That means that each element δ_{ik} of the Δ matrix represents the indices of factors satisfied in the location of the i kind of projects in the k elementary zone.

If $\delta_{ik} = 1$ the k area satisfies the demand at the required level.

If $\delta_{ik} < 1$ means that at least one demanded factor was not satisfied.

If $\delta_{ik} > 1$ the k area offers more conditions than those demanded.

The concepts of fuzzy numbers are used to evaluate mainly the subjective attributes and information related to importance of de general and specific factors.

Figure 1 presents the membership functions of the linguistic ratings, and Fig. 2 presents the membership functions for linguistic values.

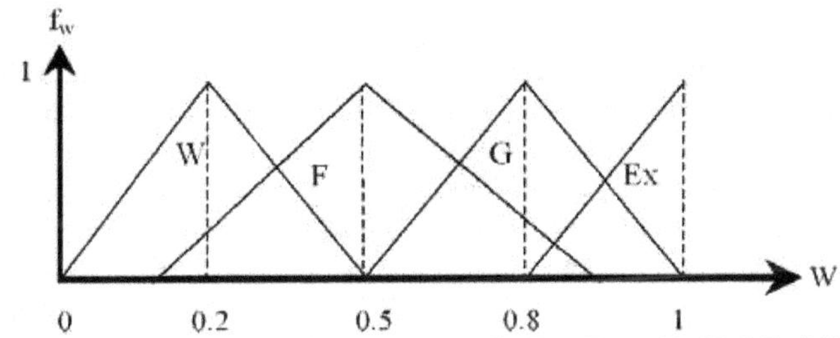

Fig. 1. Linguistic ratings: W = Weak: (0, 0.2, 0.2, 0.5), F = Fair: (0.17, 0.5, 0.5, 0.84), G = Good: (0.5, 0.8, 0.8, 1), Ex = Excellent (0.8, 1, 1, 1)

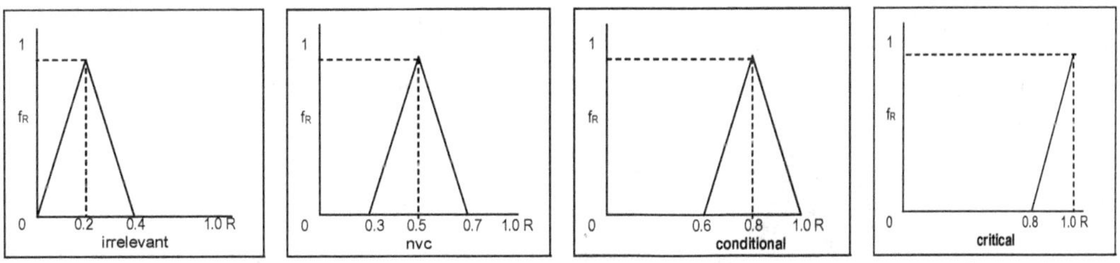

Fig. 2 Linguistic values: I = Irrelevant: (0, 0.2, 0.2, 0.4) , NVC = Not Very Conditional: (0.3, 0.5, 0.5, 0.7), C = Conditional: (0.6, 0.8, 0.8, 1.0), C = Critical : (0.8, 1.0, 1.0 , 1.0)

4. Methodology

The methodological approach consists in selecting a set of location factors that can be measured in territorial sites and associated to characteristics of under study projects. The offer and demand levels of these location factors must be defined and quantified, and a fuzzy algorithm operates the datasets obtained, in order to produce a hierarchical indication for sites and project location (Fig. 3).

The first step consists in listing appropriate location factors as resulting from territorial study and project research. Territorial study also help on site contours adopted for offer measurement, in general the suitable for available thematically data (economics, population, etc.), such as municipal or district census boundaries. Project research describe what kind and amount of facilities, resources, and logistics are necessary to improve related services and activities. The initial information is used for classifying offer and demand in several levels, corresponding to linguistic variables mentioned before in the mathematical model.

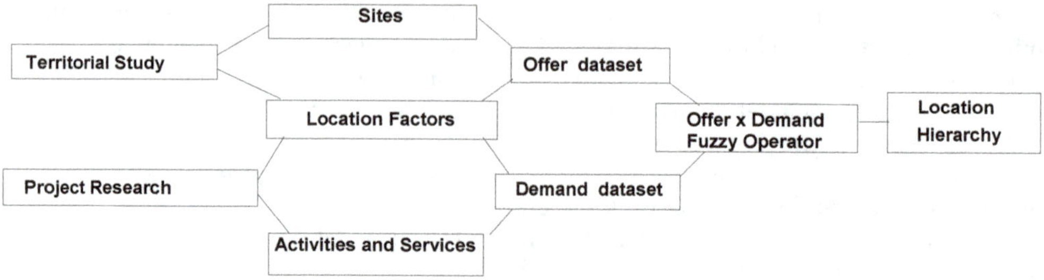

Fig. 3. Methodology

The offer is measured in levels for each considered site, and a geoprocessing tool can turn this job more effective and precise. A geographic code is used as key column for relational operations with the studied sites, as join and relates with tables containing thematic data. The number of levels can vary from 4 (four) to 10 (ten), more levels are better for classifying and displaying data in GIS ambient, but later they will must be regrouped in 4 (four) levels (Cosenza & Lima, 1991) to attempt the linguistic concept (Excellent - Good - Fair – Weak).

The rules for converting data in operational values to indicate these levels are previously defined in registry tables (relations between parameters and concepts) and could be generated by geoprocessing tools in two ways:

- Spatial analyses, when properties as distance or pertinence to georeferenced items (roads, pipelines, ports, plants, etc) are used to assign the level (Fig. 4),
- Statistic classification, when data is directly associated to the site contours (population, incomes, etc), and a range of values must be classified by statistics and grouped as assigned levels (Fig. 5).

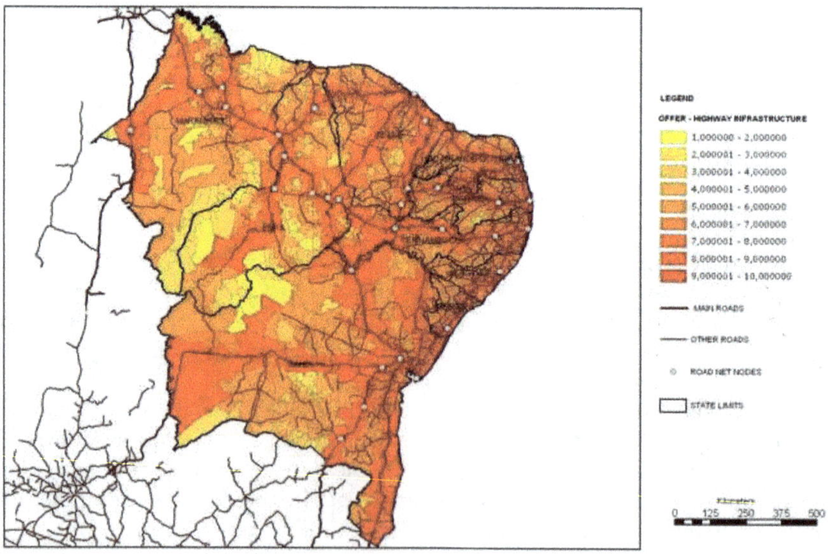

Fig. 4. Georeferenced levels of highway infrastructure offer performed by spatial analyses

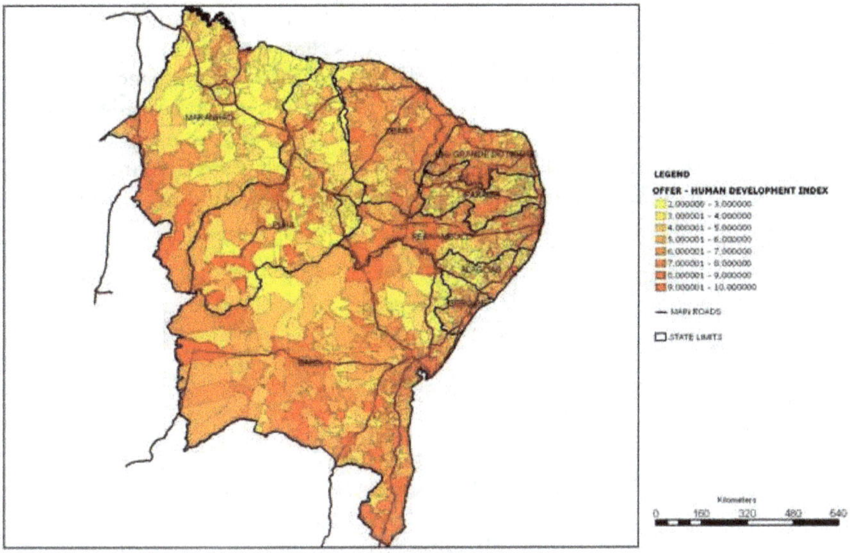

Fig. 5. Georeferenced levels of human development index offer performed by statistic classification

The demand is also organized in registry tables (Table 8), whose values are assigned by subjective interpretation of experts, based in their experience on implementing and operating similar projects. The more dependent projects are on a given factor; the highest is the demand level assignment. The demand levels can be defined in a different number them offer levels, but 4 (four) levels could deal more properly with the linguistic concept (Critical - Conditional - Not very conditional – Irrelevant).
The factors must be defined on each project as general (G) or specific (S). As seen before, a specific factor is more impacting than a general factor, because less offer of specific factor (natural resources, climate, market, etc) them requested by project could harm the location.

FACTOR ID	PROJECTS:												FACTOR DESCRIPTION
	A		B		C		D		E		F		
Fat_Infra_RODOV_3	3	G	3	G	3	G	3	G	4	G	3	G	Roadway infrastructure
Fat_Infra_FERROV	1	G	3	G	3	G	2	G	3	G	2	G	Railway infrastructure
Fat_Infra_HIDRO	1	G	3	G	3	G	2	G	3	G	1	G	River transportation
Fat_PORTO_Exportador	2	G	2	G	3	G	2	G	2	G	2	G	Exportation Ports
Fat_ETANOL	1	G	1	G	2	G	4	S	2	G	4	S	Ethanol Industrial Plants
Fat_Forn_METAÑOL	1	S	1	S	2	S	1	G	2	S	1	G	Methanol Ports

S = factor set as specific, G= factor set as general
4= Critical - 3= Conditional - 2= Not very conditional – 1= Irrelevant

Table 8. Demand table: project (identity preserved) in columns, location factors in lines

After assigned, both offer and demand datasets could be inputted as arrays and processed by computational resources, that compare offer vs. demand relations for each site and each project, in order to produce an output array containing hierarchical indicators.

To rule the process is used a relationship table (Table 9), where an equal offer vs. demand diagonal is placed with value = 1, which represents situations that offer matches demand. The other values could represent lack or excess, and may be adjusted to minimize or maximize effects around diagonal. For instance, when a project still considers sites where a little lack of offer as not critical, it could be assigned values near zero for poor offer relations, if lack of offer cancel the project, all values where offer is less than demand should be zero. In other way, when is interesting to know sites with a greater amount of offer, it could be assigned an increment for best offer relations.

			WEAK	FAIR	GOOD	EXCELLENT
		0	$\mu_{B_1}(x)$	$\mu_{B_2}(x)$	$\mu_{B_3}(x)$	$\mu_{B_4}(x)$
	0	1/n!	1/(n-1)	1/(n-2)	1/(n-3)	1/n
IRRELEVANT	$\mu_{A_1}(x)$	0	1	$1 + \mu_{B_1}(x)/n$	$1 + \mu_{B_2}(x)/n$	$1 + \mu_{B_3}(x)/n$
NOT VERY CONDITIONAL	$\mu_{A_2}(x)$	0	$\mu_{B_1}(x)/\mu_{A_2}(x)$	1	$1 + \mu_{B_1}(x)/n$	$1 + \mu_{B_2}(x)/n$
CONDITIONAL	$\mu_{A_3}(x)$	0	$\mu_{B_1}(x)/\mu_{A_3}(x)$	$\mu_{B_2}(x)/\mu_{A_3}(x)$	1	$1 + \mu_{B_1}(x)/n$
CRITICAL	$\mu_{A_4}(x)$	0	$\mu_{B_1}(x)/\mu_{A_4}(x)$	$\mu_{B_2}(x)/\mu_{A_4}(x)$	$\mu_{B_3}(x)/\mu_{A_4}(x)$	1

Table 9. Relationship table for offer vs. demand comparison and attributes: on columns weak, fair, good and excellent; on lines irrelevant, not very conditional, conditional and critical

The results are obtained as a table (Table 10), where columns are projects and lines are sites, and the obtained values express how territorial conditions match project requirements. A value normalized to 1 (one) represents the situation where both offer and demand are balanced, so location is recommended. Values greater than 1 (one) indicates that the site has more offer conditions than required, and values less than 1 (one) indicates that at least one of the factors was not attempted.

105.Assis Chateaubriand	4102000	1,043	1,044	0,876	0,945	0,966	1,094	*0,995*
106.São João do Ivaí	4125001	0,986	1,008	0,898	0,944	1,018	1,114	*0,995*
107.Rio Claro	3543907	0,973	0,94	0,906	0,986	1,019	1,143	*0,995*
108.Campo Mourão	4104303	0,996	0,978	0,874	0,986	1,006	1,126	*0,994*
109.Rolândia	4122404	0,991	0,966	0,891	0,978	1,021	1,118	*0,994*
110.Imbituva	4110102	1	0,996	0,901	0,909	1,029	1,13	*0,994*
111.Ipiranga	4110508	1	0,996	0,901	0,909	1,029	1,13	*0,994*
112.Apucarana	4101408	0,985	0,99	0,93	0,97	0,999	1,088	*0,994*
113.Santa Mariana	4123907	0,969	0,968	0,9	0,955	1,053	1,115	*0,993*
114.Mal Cândido Rondon	4114609	1,024	1,019	0,891	0,96	0,999	1,066	*0,993*
115.Lençóis Paulista	3526803	0,958	1,005	0,885	0,981	1,025	1,105	*0,993*
116.Candelária	4304200	1,003	1,073	0,9	0,949	0,98	1,053	*0,993*
117.Colina	3512001	0,983	1,001	0,874	1	0,993	1,104	*0,993*
118.Ribeirão Preto	3543402	0,915	0,953	0,89	1,001	1,029	1,166	*0,992*
119.Cruz Alta	4306106	1,001	1,005	0,888	0,961	0,979	1,116	*0,992*
120.Canoinhas	4203808	1,008	1,009	0,921	0,93	0,993	1,088	*0,992*

Table 10. Hierarchies location results for a set of municipalities, where project (identity preserved) is placed in columns, with last column shows media for all projects

Table could be now georeferenced to the sites (Fig. 6) by their geographic codes, using join or relate operations with the georeferenced tables. In the next step, location indicators are classified by statistics and displayed as chromatic conventions, in order to interpret spatial possibilities of placement. The chromatic classification for results can use various statistic methods, such as: natural breaks, equal interval, standard derivation and quantile.

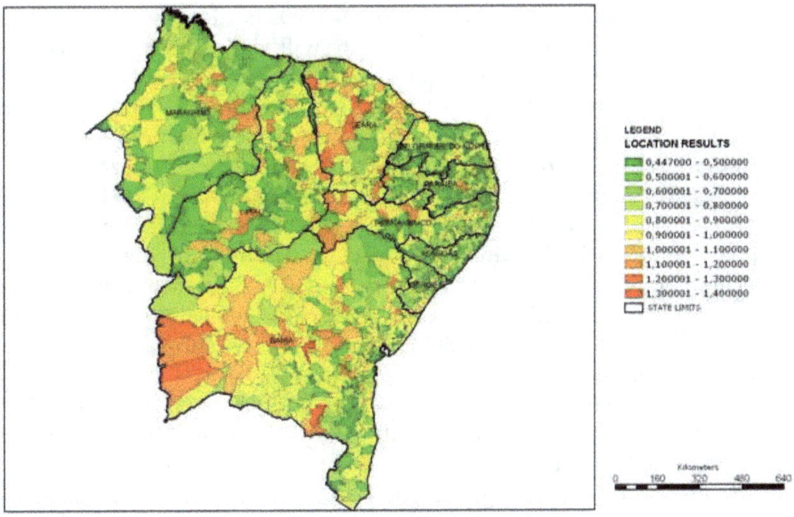

Fig. 6. Location indicators are classified and displayed as chromatic conventions

Natural breaks are indicated to group a set of values between break points that identifies a change in distribution patterns, and is the most frequent used form of visualization for identifying best location. Equal interval is used to divide the range into equal size values sub-ranges, and is used to identify results perform in comparison analysis. Standard derivation is used to indicate how a value varies from the mean, and is often used to show how results are dispersed. Quantile groups the set of values in equal number of items, and is used less frequently because results are normalized.

7. Conclusion

Location models can also be employed for previewing land use and occupation of urban areas. An analogy could be done considering an occupation typology (habitational buildings, industrial zone, etc.) as a project for an urban site (district, zone, land, etc.). A list of location factors that direct urban development could be selected from spatial, economic and social data records (population, market, education, prices, mobility, health care, etc.).

The offer of these location factors could be measured on urban sites from local surveys or official census data. Most of geographic offices in charge of registering official data make available their operational boundaries as feature classes compatible with GIS platforms.

Urban planners, engineers, public services managers, political authorities, should define the demand set, and will determinate the relevance of a factor on occupation typology, and multi criteria analysis will be helpful to equalize their opinion (Liang & Wang, 1991).

But how a location model can help urban engineering research? If a land use or activity placement could be treated as a project, ordering distinct location factors, it should be possible to measure territorial offer and typology demand. Presuming that recent placement situations can be studied to produce diagnosis based on configuration of related offer and demand sets, researching past offer sets may be interesting for understanding how factors evolution influences a site.

For instance, registering and analyzing the offer records along a significant time, and consulting specialists for demand attribute, it will be possible to isolate pattern characteristics of a situation. Observing offer increase or decrease along the time, a general urban evolution tendency (residential, industrial, commercial, etc.) could be expressed by its particular demand set. Comparing the urban site offer with a demand assigned pattern, it is possible by simulation to explore future scenarios. A georeferenced array of urban sites vs. pattern characteristics could indicate how intense each site matches the pattern characteristics, and based on the values obtained verify the possibilities of occurrence.

So, if the responsible authority inquires about a place that would be a commercial zone in the next five years, the researcher would construct an offer fuzzy set of the urban site based on recent data, and check it with a proposed pattern of typical commercial zone factors demand. The possibility of occurrence, defined by the hierarchic values, could be used to determinate and prioritize actions.

By extracting specific geodata of offer and demand sets, it is also possible to identify which factors have significant influence on the results, and so define strategic intervention that could direct the expected results.

To conclude, an offer and demand logic operator attached to geoprocessing resources could enhance the horizon of researches on urban engineering methods, and improve queries and simulations that will help to understand and simulate the dynamic of cities growth.

8. References

Attanasio, D. & alii, (1974). *Masterlli- Modelo di Assetto Territoriale e di Localizzazione Industriale*, Centro Studi Confindustria, Bologna

Attanasio, D. (1976). *Fattori de Localizzazione nell'Industria Manufatturiera*, Centro Studi Confindustria, Bologna

Cosenza, C. (1981). *A Industrial Location Model*, Working Paper, Martin Centre for Architctural and Urban Studies, Cambridge University, Cambridge

Cosenza, C. & Lima, F. (1991). Aplicação de um Modelo de Hierarquização de Potenciais de Localização no Zoneamento Industrial Metropolitano: Metodologia para mensuração de Oferta e Demanda de Fatores Locacionais, *Proceedings of V ICIE - International Congress of Industrial Engineering*, ABEPRO, Rio de Janeiro

Curry, B. & Moutinho, L. (1992). Computer Models for Site Location Decisions, *International Journal of Retail & Distribution Management.*, Vol. 20

Harlow, M. (2005). *ArcGIS Reference Documentation*, ESRI: Environmental Systems Research Institute Inc., Redlands

Jarboe, K. (1986). Location decisions on high-technology firms: A case study, *Technovation*, Vol. 4, pp. 117-129

Kahraman, C. & Dogan, L. (2003). Fuzzy Group decision-making for Facility Location Selection, *Information Sciences*, p. 157, University of California, Berkley

Liang. G. & Wang, M. (1991). A Fuzzy Multi-Criteria Decision-Making Method for Facility Site Selection. *Int. J. Prod. Res.*, Vol. 29, No. 11, pp. 2313-2330

Lima, F., Cosenza, C. & Neves, C. (2006). Estudo de Localização para as Atividades de Produção do Biodiesel da Mamona no Nordeste Empregando Sistemas de Informação Georeferenciados, *Proceedings of XI Congresso Brasileiro de Energia*, Vol. II, pp. 661-668, COPPE/UFRJ, Rio de Janeiro

2

A Contribution to Urban Transport System Analyses and Planning in Developing Countries

Giovani Manso Ávila
Federal University of Rio de Janeiro (UFRJ)
giovani@ufrj.br
Brazil

1. Introduction

The purpose of this chapter is to provide a contemporary statement of existing approaches to transport-land use planning in urban and regional areas to advise practising transport planners and students. The chapter concentrates on the knowledge that are available for the synthesis, analysis, and evaluation of alternative land use-transport-systems plans and regulations. First, the transportation planner must develop an understanding of the planning process before attempting to address herself to the broader polices issues. The first objective of this chapter is to provide transportation planners with an understanding of the land-use models and their application to urban and regional planning problems. The final objective covers policy, regulations and plans.

Absence of physical space between people and firms is the definition of Economists to cities. Cities appear to supply the necessity to eliminate transport costs. Density lowers the costs of interacting with other people and speeds the flow of goods, people and ideas. The location and structure of cities is driven by the desire to eliminate transport costs. In this context, transportation technologies have been the primary determinant of the location and structure of cities. There is increasing concern across the world about increasing traffic congestion and the costs it imposes, particularly on accessibility, the environment, other social factors, such as accidents, and the economy in general. Growing personal car-mileage is engendering the well-known effects on the environment and transport systems, especially roads, which are not able to cope with the increasing amounts of traffic. Further development of innovative, integrated and well-balanced policies is strongly needed. Strong and ongoing growth in mobility, especially in road traffic, means that transport trends are unsustainable if only current policies are pursued: with constraints on resources, space, safety and the environment there are only limited possibilities to extend transport supply to safeguard accessibility. It is quite clear that the need for travel cannot really be avoided; it is crucial for the performance of social and economic functions in any society. People are not travelling much more often than twenty years ago, but they are travelling further and with greater use of the private car. Increasing car ownership is a central component of this (Transland, 2000). There is a lack of knowledge about the interaction between land use and transport and the related planning consequences. Institutional demands for integrated policymaking is another problem. In this context, we approach in the following sections a systematic view of a transportation system planning and analysis, and its relations with land use polices, indicating the natural evolution of the cities, it's intrinsic relations with transports and the needs to intervention and control by regulations polices.

2. Phases of Urban Development

The phases of urban development has a link with the economic development of transportation. They are: (a) Initial capitalism phase; (b) Industrial development; and (c) Big traffic problem on the cities.

There is an hypothesis that the city is a spatial expression of certain economic structure and productions relations as follows: (a) Feudalism - It had only one administrative centre of the autarchic production that also acted as producing centre of small services. Space reduced with commutes in the majority by foot; (b) Mercantilism - The city starts to be a centre of commercial interchange and a centre administrator of commercial interchanges. One becomes producing centre of manufactures and services. The centre expands and foot did the commutes (distances about 5km.); and (c) Industrial Capitalism - The city becomes centre of industrial production. It has an intensification of the social division of the work producing concentration of workmanship hand and demographic growth. The city is centre of the public, private administration, of finances and advertising.

There are Advantages of agglomerations with direct communication with other producers and central offices agencies of public and private sectors: suppliers, administrative facilities, banks, competition partnership, infrastructure of services like telecommunications, light, force, post offices. Accessibility to market and workers: attraction for the diversity of services, transportation facilities. We can nominate the spatial effects of industrial capitalism as follows: (a) Big Urban Expansion; (b) Land use and occupation with variables urban functions; and (c) Development of bulk transportation to supply, manufactures, etc.

Cities in the new world have traditionally been transportation hubs. When colonisation began, the function of cities tended to be facilitating the flow of the wealth of the new world to the markets of the old world. As such, cities located either close to the sources of wealth or in natural transport hubs. Thus, Lima located near the mineral wealth of Peru. Salvador da Bahia, Rio de Janeiro and Buenos Aires were all ports. The major cities of the U.S.—New York, Boston, San Francisco and New Orleans—were all located in places where rivers meet the sea to take maximum advantage of water born transit. Even more generally, the world was dotted with smaller cities that serviced an agricultural hinterland. Because transportation costs were high, it made sense to have commercial and manufacturing centres that were physically close to the vast majority of people who worked the land. Medium sized cities dotted the landscape because they needed to be close to the farmers.

Gradually, rail networks created a reduction in transport costs and made the locations of cities more flexible. Late 19th century cities like Chicago and São Paulo grew as railroad hubs. Rail also made it possible to have fewer, bigger cities that were further from the farmland. The mechanisation of agriculture also meant that fewer people were required to work the farms, so there was less need to medium sized cities that were close to farmland.

The 20th century has seen this process continues. Trucks and improvements in rail have further reduced transport costs. The advantage of water-born transport, particularly for intra-national transport, has further disappeared. The result of this transformation can be

seen throughout the world. In the 19th century, urban areas did well if they had advantages that favoured producers. In the 21st century, urban areas will do well if they attract potential consumers. This shift has already created a massive dislocation within the USA as the colder cities of the rustbelt have been replaced by the newer edge cities of the sunbelt.
As such, national transportation infrastructure will tend to be extremely important for the changing urban landscape. Continuing improvements in transportation will certainly continue to change the growth of cities. However, as important as this process is, it does not necessitate a new, government approach to transportation policy. As long as the government continues to develop highways that are hopefully paid for by their users this process of urban change will continue, probably in a reasonably efficient manner.

2.1 Phase 1 - Initial Phase of Capitalism

Phase characterised by little industrial development, little unity of production and low demographic density. Spatially we have: (a) Production centralised in a old historic centre inner circle; (b) Disposition workmanship near productions centres; (c) Industrial location as a function of territorial supply of production factors like transportation facilities; and (d) Public and private administration and financial centre in the city centre (Fig. 1).

2.2 Phase 2 - Big Industrial Development

From the third part of century XIX, in the European cities a bigger development of the industrial production with new characteristics was observed: (a) Bigger social division of the work; (b) Concentration of the production; (c) Bigger units manufacturer; (d) Bigger concentration of workmanship and demand for mass transport; (e) Economies of scale and agglomeration; (f) Bigger volume of production and transport; (g) Expansion of the tertiary sector (wholesale trading, banks, private and public administration), necessitating of contacts right-handers and consequent agglomeration in the centre of the city; (h) Social stratification and space segregation; and (i) Development of mass transportation networks (trams with animal traction, urban railroad network, electric tram)(Fig. 2).

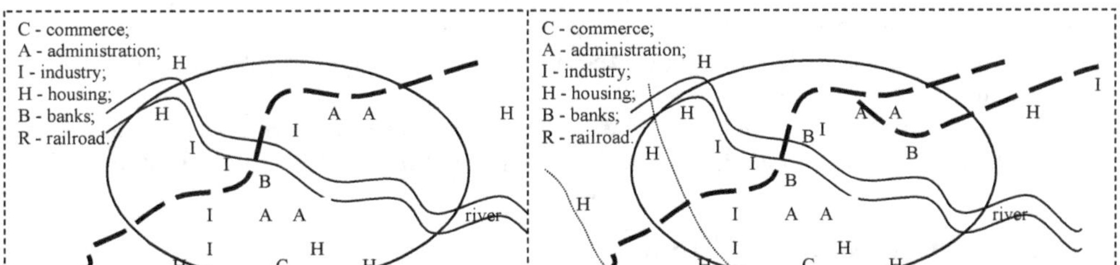

Fig. 1 and 2. Initial phase of capitalism, and big industrial development

2.3 Phase 3 - Surface Transport Development

With the growth of automobile industry, mainly after the end of 1st. World War, a new age of urban occupation appears, conditioned by the development of roadway modality, with growth of automobile consumption and, after a 2nd. World War, a growth in the buses consumption. As a consequence we have: (a) Occupation of the spaces between the transport axles; (b) Growth of the urban periphery; (c) Expansion of the road net and the demand for parking; (d) Ideal of the "motorised city" and the "proper car"; and (e) Intensification of the functional separation of land use implying in: bigger concentration of the tertiary sector in the centre, expansion of the economic functions of the tertiary, increment of the economic concentration, requirement of qualified workmanship, and demand for radial flows of transport (Fig. 3).

2.4 Phase 4 - Big Traffic Problems on the Cities

After a phase of full development of the individual transport, from middle of years 50, the cities of the first world had evidenced that: "The individual transport acted as narcotic of the functioning of the urban economic system". As a consequence we can see: (a) Growing problems of congestion and security; (b) Reduction of accessibility to the city centre; (c) Lost times in transit; (d) Demand access restriction to markets and for workers mainly to central areas; (e) Changes in land use with growing demand for urban network what cause a concurrency with tertiary sector; and (f) Hanging of central areas (Fig. 4).

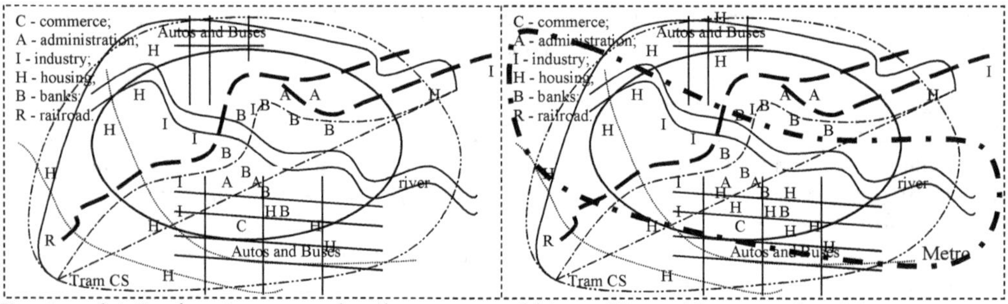

Fig. 3 and 4. Surface transport development, and big traffic problems on the cities

From this conclusions the trends was consider mass transportation as a necessity to achieve development of urban economic system. This kind of transport allows: (a) Growth of accessibility to consumption centres; (b) Growth of accessibility to workersmanships; (c) Facility to flow of goods; (d) Turn bigger the influence area of city centres; and (e) Add new regions to urban system – conurbation.

The mass transportation systems like subways (Metro) and urban surface train – tram, have a new mission and several big cities retrieve big constructions of this kind of systems.

3. Urban Centralisation and Transports

Analysing the dynamics that defines a differentiated occupation of the tertiary sector of the economy in the urban context and its consequences for the transports we can see the importance of this relations.

3.1 The Tertiary Sector of the Economy

Characterised by the service sector, provides services to general population, including: (a) Public administration and services; (b) Private administration and services; (c) Financial sector; (d) Business sector in wholesale and retail; (e) Fun and entertainment (movies, television, radio, music, theatre, etc.); (h) Health; and (i) others services.

In most developed and developing countries, a growing proportion of workers are devoted to the tertiary sector. In the U.S., tertiary workers compose more than 80% of the labour force. The Law of Land Use of a city regulates where the urban activities can take place inner the city. When allow or nor certain activities in certain quarters, the law tries to give to the city a desired occupation. If the law does not impose restrictions to the busy area for the tertiary Sector and, it does not charge regulatory taxes for the land transference and building, the free market goes to function. In these conditions the income-producing activities, as of the tertiary one they go if imposing on less income-producing, or not income-producing as the housing. Gradual the not economic activities are banished from certain centres giving place to the sector of services. This sector needs centrality and also: (a) Agglomeration advantages; (b) Economic growths; and (c) Land Use Regulation.

3.2 Concurrency Between Secondary and Tertiary Sector

The tertiary sector is more dependent of central location, while the industrial sector is more ground dependent for the mechanised production. The industrial sector can be regional, national or international, in contrast of the retailing that is local. The conclusion is that the secondary sector searches the periphery, leaving free space in the centre for the tertiary sector, whereas the elementary school is basically agricultural. In the competition for the spaces, to if more valuing the ground for the increase of the demand in the points central offices, paid the activity of bigger return, in the case the tertiary one. In such a way the housing, leisure and small commerce go being banished of the urban centres. A social problem appear when the less gains, more far and expensive is do activities of work, leisure, etc. In case that it has not measured of control of the public power this comes to be a strong incentive to the in slums process (*favelização*).

4. Spatial Dynamic of Land Use and Transportation

The better way to understand and evaluate a transportation system is to know the spatial dynamic of land use and transportation (Fig. 5).

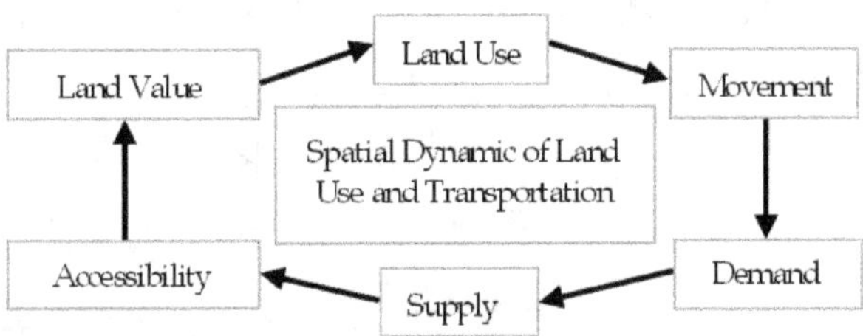

Fig. 5. Spatial dynamic of land use and transportation

The components can be described as: (a) Land Use - is understanding as a joint of activities realised, to live and work, fulfilling the necessities of food, live, generate exceeding to trade and then achieve the grade of auto-sustain. This role of activities can be classified in one or more of three sectors of the economies that will establish the dynamic of occupation and space utilisation. Each activity that takes place in an urban or rural space does demand movement that are derivatives of travel necessities to do business, work and play; (b) Movement - the movements that occur in one space are derivatives of an interrelationship between the role of complementary activities classified in: housing, work, business or a combination between them, involving bulk or passenger; (c) Transportation demand - the social economic activities get movement's necessity, which generates a demand for transportation systems until the desirable service could be reached; (d) Transportation Supply - can be viewed as a response of a physical or political environment to demand attempting and is related to initial conditions of regional supply of this facilities like rivers, sees, geography and whether, or some interventions that must be taking place by the man to provide the integration between regions and complementarities or trades; (e) Accessibility - the supply of transportation systems generate facilities to feasible in one greater scale the intensity and diversity of land use, supplying not only the initial potential but creating news opportunities of use, differing some times absolutely of which that generates them; and (f) Land Value - increasing accessibility costs trade-offs decreasing transportation costs and increasing land value, changing the role of activities and creating new land use.

This new activities will taking gradually the spaces of older that will take place in another space or disappear of this region, turning goods or services incoming from another regions, establishing a new pattern of interrelationship between regions. This characterises the spatial dynamic of land use and transportation.

In Fig. 6 we see that this cycle of activities does not end. The changes in land use will create new patterns of movements that will demand new transportation systems, changing the land value, and so on.

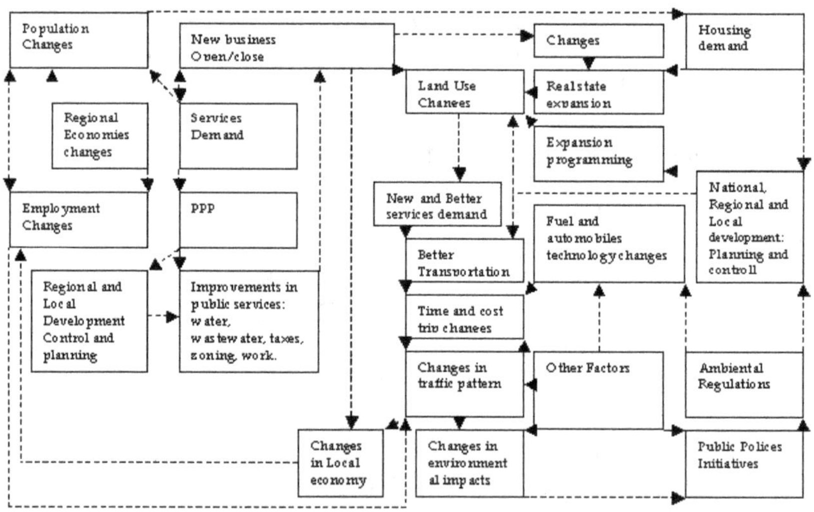

Fig. 6. Local and regional dynamic originated from relationships between transportation and land use

5. Theory of Land-Use Transport Interaction

Theories on the two-way interaction between urban land use and transport address the location and mobility responses of private actors (households, firms and travellers) to changes in the urban land use and transport system at the urban-regional level. That urban land use and transport are closely inter-linked is common wisdom among planners and the public. That the spatial separation of human activities creates the need for travel and goods transport is the underlying principle of transport analysis and forecasting. Following this principle, it is easily understood that the sub urbanisation of cities is connected with increasing spatial division of labour, and hence with ever increasing mobility. However, the reverse impact from transport to land use is less well known. There is some vague understanding that the evolution from the dense urban fabric of medieval cities, where almost all daily mobility was on foot, to the vast expansion of modern metropolitan areas with their massive volumes of regional traffic would not have been possible without the development of first the railway and later the private automobile, which has made every corner of the metropolitan area almost equally suitable as a place to live or work. However, exactly how the development of the transport system influences the location decisions of landlords, investors, firms and households is not clearly understood even by many urban planners. The recognition that trip and location decisions co-determine each other and that therefore transport and land-use planning needed to be co-ordinated led to the notion of the "land-use transport feedback cycle" (Fig. 7).

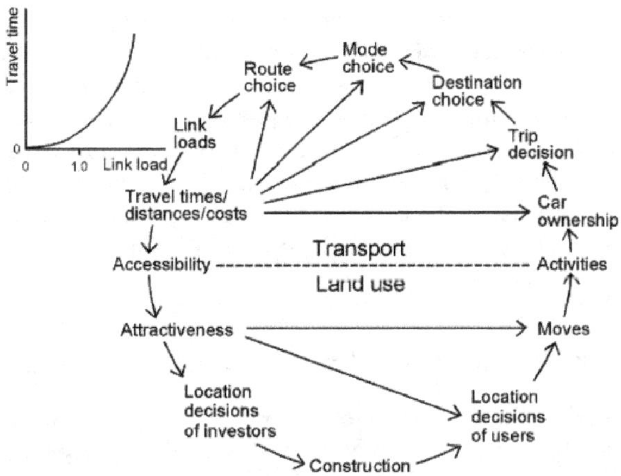

Fig. 7. The "Land-use transport feedback cycle" (Transland, 2000)

The major theoretical approaches to explain this two-way interaction of land use and transport in metropolitan areas include technical theories (urban mobility systems), economic theories (cities as markets) and social theories (society and urban space). A variety of 'ideal' land-use transport systems as optimal solutions to urban land-use and transportation problems have been formulated since the late 19th century. These systems vary with regard to spatial structure, residential density, distribution of land uses and predominant transport mode. Attempts to determine the 'ideal' land-use transport system in contemporary cities have yielded different results. While it has almost become common wisdom that systems involving dispersed development are much less favourable with regard to average trip length, energy consumption, greenhouse gas emissions and land take, there is no unequivocal evidence for the advantages of either compact-city or decentralised-concentration policies. The results of empirical studies of land-use transport interaction conclude that residential density has been shown to be inversely related to trip length (Transland, 2000).

Centralisation of employment results in longer trips, while trip lengths are shorter in areas with a balanced residents-to-workers ratio. American studies confirm that attractive neighbourhood facilities also contribute to shorter average trip lengths. The theoretical insight that distance of residential locations to employment centres is an important determinant of average trip length has been confirmed empirically. The larger a city is, the shorter are mean travel distances, with the exception of some of the largest metropolises. None of the studies reported a significant impact of any factor on trip frequency. Residential and employment density as well as large agglomeration size and rapid access to public-transport stops of a location were found to be positively correlated with the modal share of public transport. 'Traditional' neighbourhoods showed a higher share of non-car modes. Accessibility was reported to be of varying importance for different types of land uses. It is an essential location factor for retail, office and residential uses. Locations with high accessibility tend to be developed faster than other areas.

The value of accessibility to manufacturing industries varies considerably, depending mainly on the goods produced. In general, ubiquitous improvements in accessibility invoke a more dispersed spatial organisation of land uses. Regarding impacts of transport policies on transport patterns, causal relationships are relatively undisputed, and empirical studies largely agree on the impact mechanisms. While travel cost and travel time tend to have a negative impact on trip length, high accessibility of a location generates longer work and leisure trips. Studies on changes in trip frequency are only known for travel time improvements, where time savings were found to result in more trips being made. Mode choice depends on the relative attractiveness of a mode compared to all other modes. The fastest and cheapest mode is likely to have the highest modal share. However, offering public transport free of charge will not induce a significant mode switch of car drivers, rather of walkers and cyclists.

6. Review of Current Transport and Land Use Planning Issues

The review covered both technical, behavioural and institutional issues, i.e. impacts of local land-use policies on the behaviour of travellers and, vice versa, impacts of transport policies on the location behaviour of households and firms within urban regions ('What'), as well as issues of co-ordination of land use and transport policies in different national and regional institutional contexts ('How').

Urban land-use transport models incorporate the most essential processes of spatial development including land use and transport. A number of integrated land-use transport models are in use today. There are significant variations among the models as concerns overall structure, comprehensiveness, theoretical foundations, modelling techniques, dynamics, data requirements and model calibration.

The transport sub models used in current land-use transport models do not apply state-of-the-art activity-based modelling techniques but the traditional four-step travel demand model sequence (Ben-Akiva, 1974; Ben-Akiva & Lerman, 1985; and Ben-Akiva et al., 1996), which is inadequate for modelling behavioural responses to many currently applied travel demand management policies. It is a limitation.

In the future, the integration of environmental sub models for air quality, traffic noise, and land takes and biotopes are likely to play a prominent role. Issues of spatial equity and socio-economic distributions are expected to gain similar importance in model building.

Different policies affecting the location of workplaces including the construction of peripheral industrial estates and out-of-town shopping centres as well as an equal distribution of employment and population were investigated. It was found that decentralisation of facilities negatively affects the economy of the inner city while trip length and mode choice depend on the specific location and spatial configuration of population and facilities in the decentralised areas. When examining housing policies, neither the centralisation of population nor residential development in sub centres were found to have a significant impact on key transport indicators.

Land use planning policies have a major impact not only on spatial development but also on travel patterns. Development restrictions, e.g. a green belt around the city, can retard the sub urbanisation of population and workplaces thus strengthening the economy of the city centre. The construction of an outer ring road results in further decentralisation, relief of congestion and increasing travel distances. New public transport lines have little impact on location choice but tend to strengthen the inner-city economy. Introducing speed limits results in shorter trips and increased use of public transport. The effect of increased fuel taxes on the number and length of car trips is particularly strong. Significant fuel tax increases curb the further dispersal of residences and workplaces. Higher downtown parking fees generate negative economic effects in the centre and make out-of-town shopping centres more attractive. Public transport use free of charge reinforces a pattern of centralised employment and decentralised residential locations. Volume and length of car trips remain by and large unaffected by this measure. The ESTEEM study (1998) showed that the share of automotive travel in modal choice decreases with increasing size for cities above a threshold of 750,000 inhabitants. For cities below the threshold, a slightly positive relationship between city size and car use was found.

7. Transportation Planning

Urban and regional transportation planning process is very important because turn land use sustainable. The transportation planning process will generate a legislation allowing monitoring and control land use as it was planned. A sustainable development can be defined as the development that assures the satisfaction the needs of population, without jeopardising the capacity of the future generations to satisfy the own ones: (a) To assure that the standard of life (rent available) of all the inhabitants surpasses the survival threshold; (b) To assure a good quality of life to the population, as far as access to basic grants and rights (education, health, environmental quality, historical patrimony, house, etc.); (c) To assure the equality opportunities, the right to the own culture and the rest of fundamental rights of the person; and (d) To promote that the obtaining of a certain level of development for the present population does not imply to subordinate that the future inhabitants cannot accede to resemblance or better levels of development and, in particular, to assure that the natural and cultural patrimony is not reduced.

In order to grant sustainability in the development processes it is essential: Information, awareness, commitment and public participation in the fixation of objectives and activities, and in the co taking responsibility in the profit of those with the materialisation of the same. It interesting at this point present the differences between growth and development: (a) Growth - the concentration is over the quantitative increases of different social variables; and (b) Development - it implies the improvement of the "standard of life" and of the

"quality of life" of the people. Therefore, not only it incorporates aspects of quantitative nature, but essentially of qualitative nature.
There is a creation of sustainability when promoting the integration of transport and land use planning. Three main dimensions of comprehensive sustainability which cannot be seen in isolation are identified: Environment, Society and Economy. Economic efficiency is one part of the sustainable triangle (Fig. 8) and it will be influenced by integrated land use and transport patterns. The integrated approach of Transland mainly targets on creating spatial urban patterns as well as transport patterns which fit into these spatial structures in order to ensure the development of sustainability.

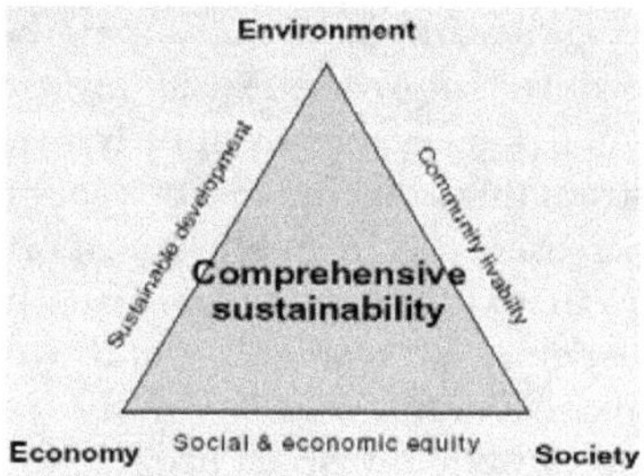

Fig. 8. Sustainable triangle (Transland, 2000)

The "Developing the citizens network" (1998) by the EU sets practical methods for making transport systems more sustainable and shifting away from excessive dependency on private car use: (a) Raising the quality and accessibility of public transport services; (b) Making walking and cycling more attractive; (c) Reducing the demand for travel, for example by reversing the trend of dispersing of functions to places which are hard to reach except by car; (d) Removing psychological barriers to the use of alternatives to cars and winning public support for policies to encourage more use of these alternatives; and (e) Making transport an essential component to strategies of spatial planning (Williams, 2005).
The integration of land use and transport planning can only provide a partial contribution to the implementation of sustainability, with impact on sustainable development within the areas of ecology, economy and society. An integrated planning approach develops structures in which ecological, social and economic sustainability can be promoted. Two basic strategic goals can be identified: (a) Land use goal - "fulfilling land use needs occupying fewer space in a better way"; (b) Transport goal - "fulfilling travel needs through environmentally friendly modes".
Sustainable mobility can be achieved considering the following chain of goals/actions: (a) Improve accessibility and the use of the space; (b) Increase the environment-friendly modes share (public transport, cycling, walking); (c) Reduce congestion; (d) Improve safety; (e) Reduce air pollution, noise, and visual nuisance; (f) Developing and maintaining a wealthy and healthy urban economy; and (g) Ensuring social equity and transport opportunities for all community sectors.
Diagnostics consists of a check-up, or evaluation of traffic and transportation existing systems including traffic generation hubs: housing, shopping centres, economic activities, etc. The planning starts with a data collection and a diagnosis of all existing transportation systems and zoning legislation. Planners will be able then to design the new pattern of land use according with the existing and future infrastructure.

8. Transportation Planning Process

A Systematic view of a transportation planning process can be explained in a simple steps as follows: (a) Determining the study area; (b) Establishing of desirable transportation and traffic conditions in the year project; (c) Determining existing studies and planning; (d) Check up of actual situation; (e) Determining actual flow pattern on the multimode transportation network; (f) Evaluation of network capacity and reserve to future projected demand flows; (g) Proposals of improvements and magnifying and implantation of transport subsystems; (h) Plan review; (i) Master plan; and (j) Indication of the financing sources. All areas of performance are described in the next section.

8.1 Transportation Planning - Areas of Performance

Urban Transportation Planning perform transportation systems conception as a hole, starting from a transportation demand forecasting and the development scenarios definitions of transportation systems. The transportation engineering performs a tools development to organise city traffic so that the planning process turn effective. The management and operation of public transportation systems perform procedures to grant the functionality of public transit systems and manager the system to achieve the desired goals and objectives reflected by operational performances of such systems. The transportation network modelling is commonly represented by graph theory and network theory for system modelling and analyses the flow distribution in a transportation network. Some examples of network analyses are: CTA Project (by Area Traffic Control - Transyt); and Transcad-GIS-GPS (http://www.apontador.com.br; Googlemaps).

8.2 Common Transit and Transportation Systems

Improvements in transportation have created a major change in the location of cities, but the connection between transport technologies and the internal structure of cities is at least as large. Traditional European towns were built for people who got around by using their feet. These cities were extremely dense. Generally homes and jobs could not be further apart than the distance that could be covered on foot. Markets also had to be small and local. The rise of public transportation permitted a change in urban form. Buses and subways still mean that people need to live in dense areas, but there can be much greater distance between home and work. After all, public transportation is a very time intensive technology. People don't want to get on a bus every time they have tea with a friend or go to the market. However, they will spend significant time periods getting to work. A typical bus or car city is a dispersed high-density city. The Brazilian *Favelas* are a perfect example of these phenomena. They are high-density areas that permit walking as a means of locomotion, but they are linked to employment by public transportation routes. Generally small buses—jitneys—permit these poorer Brazilians to get to their jobs.

American edge cities—suburbs with major employment centres—are cities designed exclusively around automobiles. These cities require not just one car in every garage, but several. Each mobile member of the household must have their own car to do anything. These cities are built at much lower densities. Driving three or four miles to the nearest grocery store is not, after all, a hardship. At their best, they offer lower density living with quick access to jobs and shops on relatively empty roads. While many academics find the suburban lifestyle sterile, there is no question that consumers who can afford it appear to greatly enjoy its many amenities. Within the U.S., the walking cities of the 19th century were gradually replaced by the public transportation cities of the early 20th century. By 1900, less than 7% of Americans used public transportation to get to work. Since 1950, America has seen a dramatic sub urbanisation of first people and then jobs. The typical job is now far from the city centre and the typical person lives even further out. As result, Americans consume unbelievably large amounts of housing relative to almost any other country.

Why don't more people in the U.S. use public transportation? Public transportation, despite its widespread availability in many cities, is used only by the poorest Americans outside of a few large cities. The reason for this is that public transportation is an extremely expensive technology for the average user—when cost is measured properly, including the opportunity cost of time. Commuting times for public transport users are much higher than commuting times for drivers. This time cost comes primarily from the fixed time cost of public transportation—this is the cost of getting to the pickup spot, waiting for the bus or train, and getting from the drop-off spot to the final destination.

Few other countries have fully followed the American example, although Canada and Australia probably come closest. European countries have massively taxed gasoline and massively subsidised public transportation. The impact of this has been to stop European cities from evolving towards car-oriented places. Latin American cities have not fully followed the U.S. model because automobiles remain too expensive for the vast majority of citizens. Will Brazil move towards the American edge city model? It seems likely that Brazil will continue to get richer. If this process continues then it seems almost inevitable that car ownership will rise significantly and urban land use patterns will start to come closer to the U.S. model. There are two potential barriers to this transformation: increasing gas prices and government regulation. Some experts believe that increasing use of fossil fuels will push the price of gasoline up many times. The historical record suggests that high prices will tend to create striking technological responses. In the short run, higher prices will be offset by conservation technologies (more efficient car engines). In the long run, higher gas prices will be offset by alternative fuels with can also power cars (like ethanol in Brazil). The probable hypothesis is that cars will remain cost effective even as fossil fuels get used up. Government regulation is of greater concern, especially in the short run. European countries have created an entirely different urban landscape than the U.S. through their different gas taxes and public transport policy. Brazil can, in principle, follow this course and keep cities dense and focused on buses, surface light rail (tram and streetcar) and subways.

One possibility is put taxes on certain types of drivers but a biased government policy against cars and car cities is contrary to the principles of economics. While the government certainly has no obligation to subsidise the car, economics tells us that consumers are better judges of what makes them happy than governments. Even if some urban analysts dislike the world of suburbs, it seems like an outrageous piece of governmental restriction on freedom to try to deny consumers their ability to choose how to live and how to commute.

To planning the general network that will supply the necessities for desired and planned land uses pattern, for a determined region, the planners usually utilises a methodological well know tool called Four Step Method: (a) *Trip Generation* - provides the linkage between land use and travel patterns. Existing land use and travel are linked utilising techniques such as cross-classifications, trip rates or regression analysis. These relationships are then applied to estimate future travel based on the forecasted change in land use; (b) *Trip Distribution* - is the process of distributing the trips generated in each zone to all the possible destination zones available. As in trip generation, there are several types of models for accomplishing this: growth factor models, intervening-opportunity models, and gravity models. In the gravity model, the number of trips between two zones is directly proportional to the product of the number of trips produced in one zone and attracted in the other, and inversely proportional to the degree of separation between the two zones, represented as a function of travel times; (c) *Modal Split* - is the process of assigning person-trips to available modes of transportation. There are three major factors that need to be considered in this analysis: characteristics of the traveller, characteristics of the trip, and characteristics of the transportation systems. The types of techniques that have been used in the development of modal-split models include regression analysis, diversion curves, and cross-classification. A different modelling approach is also utilised that consider the probability that an individual will chose a particular alternative is a function of the characteristics of the individual and of the overall desirability of the chosen alternative relative to all other alternatives, and (d) *Traffic Assignment* - constitutes assigning the distributed volumes of trips, by mode, to individual network links. The basis for this assignment procedure is that the choice of rout is basically a decision to minimise total travel time through a transportation network. There are several techniques that can be used in the assignment procedure: minimum path, minimum path with capacity restraint, multiroute probabilistic assignment. All have this basis for this operation.

8.3 Desirable Transportation System Characteristics

Starting from the zoning plan is possible to determine all desirable characteristics to the news transportation systems that must be supplied in each stage of the development process as: (a) Road Hierarchy - road system must be designed in agreement with the road hierarchy, establishing curbs, grades and width in accordance of designed vehicle to each urban and rural area geometry (lane width, parking designs, bikeways, sidewalks cross walkers, structural, arterials, collectors and local streets). A transportation system must be put together with another systems and designed itself in a hierarchical way (Green Book, 2004); (b) Mobility/accessibility - this structure contrast mobility with accessibility and has inverse interrelationship. Local systems must be high accessibility and low mobility wit low speeds, collecting and distributing people and goods to arterials systems that must have low accessibility and high mobility. Street networks connecting the remain of the city by collectors and arterials, considering distances between them like: Arterials major – 6 km, Arterials minors – 2 km, Collectors – 1 km, and Locals – 100 to 200m. This systems must preserve the roads connectivity providing binaries systems always as possible; (c) Inter modality and terminals integration - the systems must be integrated itself, across common terminals or at least closets ones, to perform a high utilisation of each vocation and capacity; (d) Traffic Calming in local areas - the people must walk safe, comfortable, without noise; and (e) Grade separation evaluation in all collector-arterials intersections - the lost time must be minimised with the adequate design and continuous evaluation of grade separations in each intersection.

9. Transportation Demand Forecasting

Knew as allocation models: (a) Network modelling using graph theory; (b) Determine the shortest paths; (c) Network traffic loading; (d) All or Nothing allocation model; (e) Stochastic models; (f) Capacity constrained models; and (g) Network equilibrium models.

These models are very complex and utilise mathematics techniques to mitigate a system optimised traffic pattern in the entire network. The allocation process needs data like origin/destination by transportation modality. This flow separation by different modes is performed with use of modal split techniques. This kind of models consider characteristics of transportation modals and vehicles, and user characteristics like: trip motivation, income, age, sex, etc. Then is possible estimate how transportation flows split between modes. The results are then loaded into matrices origin/destination O-D. These matrices are estimated by trip distribution models utilisation.
This stage require data like number of trips living each zone an arriving in each zone, well like trip cost, measured by time, distance, etc., between each pair of centroids. Trip generation is defined as the estimation of a number of people living and entering a specific traffic zone by interval of the day and trip motivation. In order to build these models is necessary information about socio-economic level of population and what kind of activities these people realises. Is also necessary to get demographic data and to perform one analyses of distribution of peoples like where they live, work, play, buy, etc. Then, is necessary to perform a study of land use and, urban activities and determine how and where the urban equipment are located: schools, work, shopping, others.
Figure 9 shows the hierarchy of studies levels to approach transportation problems. So, considering local problems, the problem abroad is inner a greater system. When approach a local problem is important to see that this problem can be inserted in a bigger one and then, must be analysed in a major level of planning.

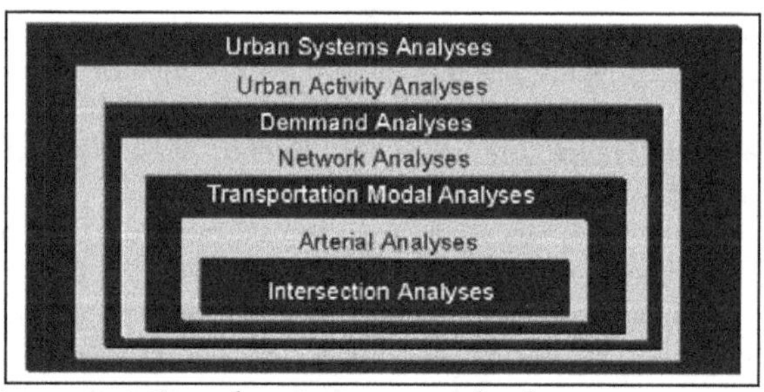

Fig. 9. Hierarchy of transportation planning system check-up

10. A Traffic Problem

The traffic problems must be approached, for example, as a problem that could be solved with local interventions, like: (a) Signalisation on street; (b) Light signals studies; (c) Traffic and pedestrian counting; (d) Phases definitions; and (e) Phases calculating (Fig 10).
The study can indicate physical changes like changes in sidewalk design or the necessity of grade separations intersections, separating flows at different levels. The problem analyses can indicate that local changes are ineffective to the problem solution and that is necessary an arterial intervention that can include: (a) Parking control at long of way; (b) Control access; (c) Speed limit control and green wave studies for transit; (d) Physical changes in arterial level; (e) Turn arterial cross section greater or smallest; (f) Duplication; and (g) Different levels intersections (Fig. 11).

Fig. 10 and 11. Node from an arterial, and arterial presented as a set of nodes and links

The problems can be greater in hierarchy like an intervention in arterial system couldn't solve the problems by itself, being necessary to perform a network analyses. In this case the traffic pattern on network is studied, and the network interventions must be done.
The urban network analyses start by defining the study region that is then divides in traffic zones (Fig. 12). This is necessary because is impossible to approach a continuous region but divided in traffic zones to be able to apply the well know models. After zoning, a hub is attributed in each zone. The models consider that each traffic is generated/attracted in hubs. The data base must contain points of origin/destination of passenger cars and transit, network design, volume, capacity and link controls, bus lines and its stops, frequency and vehicle capacity, the model is applied to estimate the pattern of traffic flow over the network, between each origin/destination hubs. By the results the annalist must determine: (a) Street changes; (b) Changes in bus lines; (c) Exclusive and priority lanes; (d) Building of bridges, tunnel and other facilities; and (e) Metropolitan subway construction/expansion. Figure 13 shows the macro area to be analysed.

Fig. 12 and 13. Zone and original arterial, and macro area to be analysed

11. Transportation Demand Analyses in Urban Areas

In Transportation Planning is very important consider that: "The transportation is a mean activity utilised to achieve an end activity". The trips are realised to enable the realisation of activities in a space separated equipment. The main question is to assure the realisations of activities. The transportation and urban systems must, in an integrated way, services to this goal. It can be achieved with an integrated urban planning that must include: (a) Land Use Planning; (b) The acts over urban activities; and (c) The acts over the transportation systems. There is a natural market failure in the transport sector. Individuals who commute to work don't internalise the effect that their commuting decision will have on other commuters. Every driver imposes a cost on every other driver. This means that too many people use the roads especially during peak hours. As long as roads are publicly managed (which is not necessary—private roads are a real possibility), this calls for a policy response. The best policy responses to the congestion externality all use the price system. Because the market failure is that drivers don't pay for the congestion they create, the best policy response will be to create tolls, or other charges that make them pay for this congestion. One price system response to congestion is a standard road toll. Ideally, these tolls will differ by time of day to reflect the different level of congestion on the road over time. Modern transponder technology means that highway tolls can be collected quickly and efficiently. In dense city streets, highway tolls will be harder to collect. In these cases, an approach like cordon pricing is generally more effective. Cordon pricing works by charging drivers to use city streets during peak time periods. One way of implementing this technology is to make drivers pay on a monthly basis for the privilege of using city streets during peak hours. Drivers then display a sticker in their window to show that they have paid the toll, and drivers caught without this sticker must pay a fee. This type of cordon pricing has been used in Singapore and elsewhere effectively. Non-price controls are almost always much less effective and more costly socially. An example is the control based in car license that restrict some cars from driving on some days. License plate numbers are used as a means of determining who is allowed to drive on which day. These proposals are inefficient because they imply that a major part of the automobile stock must lay idle for one day. Furthermore, they don't allow the people who would particularly want to drive on that day to drive even if those drivers would happily pay for the social cost of their driving. For much of the population, these car-based approaches are fairly irrelevant since they use public transportation. For this group, the key to faster commute times is improving public transportation. In general, the economic literature on public transportation has been quite clear. Buses are much more efficient than trains or subways for intra-city transport. Instead of extremely expensive extensions to the subway system, minor subsidisation of the bus network will reap much more beneficial results. In cases where traffic is extreme, it may even make sense to build tunnels for buses to drive under ground. In general, subways are almost never cost effective and sold to the public on the bases of vastly over-inflated rider ship estimates. They are particularly inefficient for cities like São Paulo or Rio de Janeiro.

12. Transports and Urban Activities - The Demand for Density

The defining characteristic of cities is density—the physical proximity of people. People come to cities and pay the higher costs for urban land because they want to be close to other people, or to other resources in the city. Economists think of the advantages of cities as coming from the elimination of transport costs for goods, people and ideas.

Physical proximity facilitates the interaction of economic actors. As such, the location and structure of cities is intimately linked to transportation technologies. The growth and decline of cities over time tends to be closely linked to changes in transportation technologies. Over the past 100 years there has been a massive improvement in transportation technologies that have greatly changed the urban landscape. In this section, I review the impact of improvements in transportation on the location and structure of cities. First, I review the impact of improvements in inter-urban transport technologies. Second, I review the impact of changes in intra-urban transportation. The inherent desire and need to perform different activities at different places implies a need for travel in any society.

The crucial planning challenge is to arrive at an optimal spatial organisation of activities (maximising opportunities) and a well balanced transport network linking these activities in an efficient and sustainable way. Therefore, land use and transport planning are highly related by nature. Finding the right balance is a delicate task in urban areas especially, with their complex activity patterns and their evident spatial and environmental constraints. Travel patterns of persons and goods are the results of equilibrium between preferences of people and companies (travel demand) and the travel conditions resulting from the supply of transport facilities and spatial patterns of activities. These preferences and conditions will determine the travel choices with respect to trip distances (distribution), mode choice and time of travel. The relation between transports and urban activities measures the specific activities and equipment distribution. Considering the activities versus required equipment, we have: (a) Work; (b) Studies; (c) Shopping; (d) Private questions; (e) Business; and (f) Play. The distribution of land uses (residential, industrial or commercial) over the urban area determines the locations of human activities such as living, working, shopping, education or leisure. The distribution of human activities in space requires spatial interactions or trips in the transport system to overcome the distance between the locations of activities.

The distribution of infrastructure in the transport system creates opportunities for spatial interactions and can be measured as accessibility. The distribution of accessibility in space co-determines location decisions and so results in changes of the land-use system.

Considering commuting as a mean to an end, transportation facilities can be viewed from the logic presented in Fig. 14.

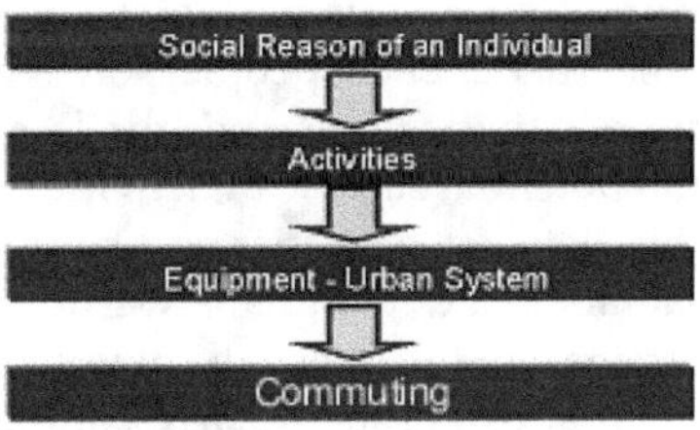

Fig. 14. Transportation facilities as a mean to an end

The society attributes to each person one status express by main activity like work, study, etc. With this attribution people have the alternative run a set of activities. By activity, we understands that's, related with urban equipment. Live, work, buy, etc., are sets of activities realised in appropriated locals like farmers, offices, industries, universities, that here are calling urban equipment. The equipment, proper to different activities, are dispersed in the urban area, and separated by variables distances. Transportation vehicles cover such distances in general. The change of activity, during the day, implies in change of the equipment and commute, covering a distance between both. The necessity of play an activity is the cause and the commute is the effect of activity played.

The social reason can be defined from the following social characteristics: (a) Age; (b) Sex; (c) Occupation; (d) Income Level; and (e) Qualification Grade. In Transportation Demand Analyses are important the characteristics that will impact individuals activities pattern and them in the number of commutes. The social and cultural environment must be taken in account. As example, in the most cultural developed countries a lower class of people goes more to theatre than in development ones. Table 1 shows an example of potential activities distribution of each social group.

Activity / Group	Work	Studies	Shopping	Privete	Business	Fun
Workers	X			X	X	X
Studants		X				X
In House Workers			X	X		X
Pensioners			X	X		X

Table 1. Example of potential activities distribution of each social group

For transportation demand analyses is important the following characteristics of activities that turn possible deduct demands forecasting of urban transportation: (a) The kind of activity will define the equipment used and then the place or city zone that will be the destination of one displacement; (b) The moment that one activity starts will define the end of a trip; and (c) The duration of such activity will define the starts of back trip.

The sets of activities that one can do, function of your status, is an ideal pattern or potential but, the supply of equipment and transportation that city can supply will imply in constrains in the standard of activities. There is constrains imposes to a sample of activities of each people, it is function of the distribution of equipment and because transportation systems. Each people have stock of time, money, and credit, physical, psychological, etc. Each activity imply in a consumption of this stocks to run the activities and to get the displacement necessary between them. Each people solve your diary schedule problem following the rule: "Realise main activities first and, the less important activities with the surplus stocks". In this way, the potential sample of activities is reduced to a real sample.

Considering the interrelationship between activities and equipment distribution in an urban area, we conclude that a special fact in a demand analysis is that living is the basis of urban activities. It is the origin and destination in almost all commutes.

Table 2 shows that home is the origin of 15,826 commutes. Then, to estimate the transportation demand is important the knowledge of equipment distribution with respect a living areas: relation home-work-home. For example, in the city of Rio de Janeiro this relation is about 97.4% (PDTU, 2001).

O/D	Home	Work	Studies	Business	Fun	Private	School	Other	Total
Home		3,430	2,727	365	504	283	347	320	7,976
Work	3,308	66	67	17	13	8	150	9	3,638
Studies	2,730	26	42	5	16	4	137	15	2,975
Business	366	3	1	5	12	7	12	6	412
Fun	516	5	1	5	21	4	20	5	577
Private	268	9	2	7	5	11	25	5	332
School	357	154	136	5	18	22	42	28	762
Other	305	7	8	6	13	3	30	8	386
Total	7,850	3,700	2,984	415	602	342	763	396	17,052

Table 2. Matrix with number of daily commutes generated from equipment of each row in the matrix and with destination to equipment in each column

Also, it is important to analyse each particular activity, the distribution of equipment to your utilisation and the characteristics that will impact the transportation demand: (a) Work - if a person works, this is a mainly activity to him. Characterised by economically active people. Highly stability in time with well knows duration, originating pendulum demand in transportation systems, with two peaks a day, in the morning and in the afternoon. There is a functional division in land use. The secondary sector is located in the border and the tertiary sector in the centre. This classification is fundamental to estimate where some activity will take place; (b) Study - mainly to a class of population. Must have a near distribution house-schools, mainly in the first grades. The last grades can be centralised. It can be classified in: Primary, Secondary, Technical, Graduation, and Post-graduation; (c) Shopping - diary supply of general goods. Housewives in general have this responsibility, which can be spread in diary, weekly, monthly. The equipment to daily shops are delivered around the city overall. This contributes to commute by walk. The equipment to buy goods at meddle and long run are sparse and located in the centre of the cities. They demand transportation systems in commutes. The distribution of equipment to diary buyers is function of several factors, like: structure of the city, population density, age of neighbourhood, and social structure of demand; (d) Private questions - several activities related with tertiary sector. No working or shopping include in this category; (e) Public services; (f) Private services - located generally in centre areas; (g) Business - activity interrelated with the job. In general is inside tertiary sector. Have a trend to be located in the central areas. Consist of private or public administration; and (h) Fun - activity played at a free time. Theoretically all peoples have time to play fun activities. The kind of fun depends of status (level of income), local culture, etc. Supply equipment is determinant. Cultural factors are determinant to intensity of activity. It is important to transportation differentiate the weekend fun from in the others days of a week, because the standards are different.

13. Characteristics of Urban Structure and Plans

There has been a remarkable revolution in transportation technologies. In the early 20th century, inter-city transportation was still very much dominated by water born transit. Gradually rail and then trucks has substantially reduced the costs of moving goods across space. These costs have also eliminated the advantages of locating near natural resources, including water and farmland. As a result, cities built around production have gradually been replaced by cities built around consumer and political advantages. In the early 20th century, intra-city transportation was dominated by walking. Over the 20th century, first rail, then buses and cars have gradually come to replace human legs as the primary form of locomotion within cities. As a result, the walking cities of 1900 have gradually been replaced by driving cities of today. These rapid structural changes have created major policy problems. People have left agricultural areas and crowded into cities. In those cities, they increasingly use cars and buses instead of their feet. As a result, traffic congestion has grown more and more sever. In Brazil, commuting times in the largest cities frequently exceed more than two hours annually. This implies that there is a huge share of national resources being allocated towards the process of getting to work and getting around the city.

The Managing Plan appears of the necessity of auto-sustainable development of the city, and as legal necessity (Constitution). To analyse existing conditions and trends; to idealise future conditions; to delineate politics and lines of action for implementation. A managing plan identifies and analyses Inter-relations between obligator and optional elements. They are obligator elements: transports, housing, economy, environment, public land use, spaces, and natural areas of protection, threats and agricultural and industrial areas. They are optional elements: urban project, public security, and cultural resources. There are some elements necessary for the preparation of the Managing Plan, like: (a) Existing land use - residential, commercial, industrial, institutional, public spaces, strips of land, agricultural area; (b) Transports - urban mesh, capacity of the mesh, volumes of traffic, offers of areas of parking and demand, stations of public transport for way, bikeways networks, sidewalk networks, maps, plants, networks of public services population and job, local economy; and (c) Special topics - historic site, archaeological, farms, etc.

The element land use in the Managing Plan presents the general distribution, localisation and characteristics of the land use present and future. Maps of future land use: they present the supply to infrastructure necessities. The element transport in the Managing Plan presents the circulation of the traffic, transit, bikeways, ports, airports, tourist railroads, routes, lines of desire of pedestrians and parking. The proposals for the systems of municipal transport must consider the plans of state and federal transport and must be integrated in these systems. It evaluation offers existing facilities and the future adequacy in capacity terms and modality, considering the regional characteristics. Plans of transports are originated from the Managing Plan and developed using the knowledge of the process of transportation planning. Systems of effective transports are basic for the maintenance of the productivity, comfort, health and security of communities and regions. A plan of transports establishes a chronogram physicist-financier to provide mobility, accessibility, security, economy and necessary quality of life to community.

There are some reasons to elaborate a plan of transports: (a) Management of existing systems; (b) Continuity of investments in existing projects; (c) Existing Realignment of services; (d) Introduction of new services; (e) Construction of new facilities; and (f) Identification of the sources of financing of existing systems and improvements.

The plans of transports vary in boarding, content and target in accordance with the space and the time that treat. Five basic types exist: Federal plans, State plans, Metropolitan plans, Municipal plans, and Zone plans. These plans must involve the: Ministry of Transports, State secretariat of Transports, Metropolitans Secretariat of Transports, City department of Transports, Representatives of the Confederations of Public Transporters, Regulating agencies, and Representatives of the communities.

Six steps exist in a development of the plan of transports: evaluation of the capacity of the system, deficiencies and necessities, establishment of goals and objectives, to define and to prioritise future necessities, to develop, to evaluate and to select potential solutions, to prepare and to adopt the plan, including public revision and comments, and finally to implement, to monitor and to evaluate the performance of the plan.

The process of planning can be condensed in four elements of a backward-fed cycle: (a) Planning of system - demand for trips evaluation of market, differentiation of market planning of service, allocation of trips/modal choice; (b) Planning of service - frequency, capacity, quality; (c) Implementation of service - fleet of vehicles; and (d) Periodic revision. Legal aspects must be considered, such as: legal legislation of existing land use, law of zoning, Constitutional law. The urban planning and the subsequent process of intervention must consider the systemic interrelationships and the hierarchy of the components.

The relations between transport and land use generates a process that must be monitored and controlled. All solutions must be verified, the foreseen impacts and its minimised effect. Users must always be the targets of a plan. Users' attention must be raised to make them interested in working for their district, to accept disturbance which cannot be absolutely avoided in densely mixed used areas, and to be prepared to reduce travelling by car. Network of citizens associations are a good mean of public participation. Public involvement, when properly raised and organised, is a viable and desirable addition to the decision making and implementation process. Preliminary studies can help the project.

A large-scale urban project should be managed on a long-term basis and the contracting authorities must play a leading role in deciding what work has to be carried out by the partnership responsible for implementing the project and then assessing the actions according to a clearly defined framework. It is essential that at the start of the project, the authorities carry out preliminary studies to effectively define all aspects of the project.

In developing large-scale developments, it is important that local authorities work as part of a 'development team'. Therefore, all professionals (architects, developers and planners) should work together and consult with each other.

The end result should be for the general public to be able to live with the different professional viewpoints and therefore for the views not to be extreme. In planning a large-scale development, even the early inhabitants should be offered the most important infrastructures for supply, education, leisure, and public transport connection in order to give them the immediate possibility to reduce their need to travel by car. The concept of sustainability as well as ecology is feasible in a large context such as building a new district.

14. Land Use and Transport Polices

The impact of high residential density in reducing average trip length is likely to be minimal in the absence of travel cost increases, whereas a high density of employment is positively correlated with average trip length. Attractive neighbourhood facilities can be seen as a 'pull' factor for reducing trip length. Since more peripheral locations usually have longer trips, trip length can be expected to be negatively correlated with city size. Little or no impact on trip frequency is to be expected from land-use policies according to the theory of fixed travel budgets. Residential and employment density as well as large agglomeration size and good public-transport accessibility of a location tend to be positively correlated with the modal share of public transport, while neighbourhood design and a mixture of workplaces and residences with shorter trips are likely to have a positive impact on the share of cycling and walking. The impact of transport on land use is mediated by a change in the accessibility of a location. Higher accessibility increases the attractiveness of a location for all types of land uses thus influencing the direction of new urban development. If, however, accessibility in an entire city is increased, it will result in a more dispersed settlement structure. The impacts of transport policies on transport patterns are clearer and stronger compared to the interplay of land use and transport. While travel cost and travel time have a negative impact on both trip length and trip frequency, accessibility has a positive impact on both. Mode choice depends on the relative attractiveness of a mode compared to all other modes. The fastest and cheapest mode is likely to have the highest modal share. In general, the theoretical considerations support the conclusion that the impact of 'pull' measures, i.e. land use measures, is much weaker than the impact of 'push' measures, i.e. increases in travel time, travel cost etc.

Due to their complementary effects, land-use and transport policies need to be combined. The combination is important to achieve synergetic effects. This refers to the relationship of investment and services and planning on the one hand and regulation, pricing and to a certain extent information on the other hand. Planning and investment policies are the most important means to reduce the need for travel, because they influence land-use and transport and represent an important pre-condition for reducing travel distances and land take and making efficient use of the transport infrastructure (pull effect). Their successful implementation is only possible if additional pricing and regulatory policies create the necessary incentives for these changes in behaviour (push effect).

Co-ordination and integration of transport and land-use planning contribute to reducing the need for travel and making the remaining traffic sustainable in the institutional and policy dimension. Co-ordination includes institutional potentials, and integration encompasses policy-related potentials. Policies in the area of land use aim at reducing the need for travel, as they primarily affect urban development and land use and therefore contribute to a reduction of distances. Transport policies aim at making the remaining traffic sustainable, as they primarily influence the travel behaviour and energy efficiency of transport. However, this analytical distinction becomes somewhat blurred in practise, especially in the long run, because transport policies as a secondary effect influence land use as well, for instance where pricing policies such as fuel taxes increase the tendency that people move closer to their work places or to the urban centre to save travel costs. There are several regulations instruments for land use and transportation. Restrictions of access to the centre and creation of radial systems of collective transport of mass: (a) Solution adopted in the Europe in years 50 and 60 and later in Latin America, following the same logic of preservation of the tertiary sector in the economy; (b) Reduced trip length; (c) Reduced trip times; (d) Increased public transport patronage; (e) Increased multimode trips; (f) Increased ratio of off-peak to peak travel; (g) Increased incidence of cycling/walking; (h) Reduced rate of increase of passenger km travelled on roads; (i) Reduction in energy used by the transport sector; (j) Reduced emissions from vehicles; (k) Reduced noise exposure; (l) Capitalisation in home/office values; (m) Attraction of new business; (n) Employment; and (o) User/citizens satisfaction.

14.1 Planning Co-ordination

Planning co-ordination is complex due to the different relevant spatial levels (local, regional, national), the different sector policy fields involved and the variety of public and private actors and stakeholders involved. Institutional barriers and insufficient planning regulation and procedures are likely to be partly responsible for the uncoordinated and fragmented current planning practice. Further deregulation and liberalisation can intensify this co-ordination problem. The momentum of market forces will become increasingly important, and will impel a fundamental consideration about the future role and legitimacy of pro-active and re-active public planning. The World can play a stimulating role in the development and implementation of policies in this promising field of integrated land-use/transportation planning. Such a role calls for the incorporation of effective and feasible planning concepts and institutional innovations, and further harmonisation in the field of planning regulations and procedures.

14.2 Policies for a Better Integration of Land-Use and Transport Planning

As far as policy-related potentials are concerned, policies to better integrate land use and transport to reduce the need for travel and to make the remaining traffic sustainable were identified. The different policies were assigned to the following policy types: investment and services, planning, regulation, pricing, and information and informal policies.

When people walked to work, congestion was an issue, but it was a much smaller issue than car-created congestion. Cars take up so much more road space than consumers—both directly and indirectly through the distance needed between cars—that inevitably the introduction of cars created a massive increase in the problem of congestion.

The congestion problem is one of the most classic examples of a market failure, or externality in economics. The basic problem is that the time cost of commuting per commuter is increasing with the number of commuters. When any individual decides to drive to work, that individual takes into account the time cost of commuting to himself. However, this individual does not internalise the time cost that his commuting imposes on others. Put formally, the total time cost of commuting equals the number of drivers multiplied by time per driver, which is itself a function of the number of drivers. The private time cost faced by each driver equals just the time per driver. However, the social cost imposed by each driver equals the private cost (or time per driver) times the number of drivers times the change in time created by an extra driver. This second component is the externality, and it can be quite large. While the change in average commuting time created by a single extra driver is itself quite small, this change in average commute times must be multiplied by the overall number of drivers, which is an extremely large number. The product of these two numbers can be quite large; indeed the externality can larger than the private cost paid by drivers. The classic economic response to this type of externality is a tax on drivers equal to the externality, i.e. the number of current drivers times the marginal cost of extra time created by a new driver. In practice, this value may be difficult to estimate, and the size of tolls is frequently limited more by political considerations than by economics.

Nevertheless, the goal suggested by economics is to get people to internalise the social costs of their actions through some pricing mechanism. Once this goal is clear, then it is obvious that general taxes on driving will be extremely inefficient means of achieving this end. For example, gasoline taxes are often vaunted as one approach to congestion. But these taxes are too broad brush and often have unforeseen consequences. There is no reason for a traffic tax on cars driving on empty country roads. There is no reason to charge drivers for using city streets at 2 a.m. when there are no other cars on the road. Ideally, road taxes will be targeted so that people only pay when their driving increases congestion. The principle of targeting is important because it encourages substitution out of high congestion times and places into low congestion times and places. If all driving is taxed equally, as it is by a gas tax, then the tax does not encourage people to commute at off-peak hours rather than peak hours. However, much of the gains from commuting taxes will come from this type of inter-temporal substitution. This it is particularly vital that taxes be tied to particular times of day, or better yet road usage.

14.3 Implementing Road Taxes

As a practical matter, there are really two different types of roads, which need to have congestion taxes. The simpler situation occurs when there are clear highways that can have tolls. These tolls can no be paid through a car-based transponder which will directly by tied to a credit card. Alternatively, individuals may pre-purchase credit on the transponder. In either case, the transponder works electronically with almost no need for the driver to slow down. As a result road tolls can occur with little reduction in road velocity.

Road tolls should ideally change based on road usage. In principle, there could be a fixed toll schedule that rises during peak hours of driving. It is important for the changes over time not to be too discrete, so there doesn't develop a bulge of road use immediately before fees go up. An even more sophisticated approach is to have the fee directly tied to current road use. This fee would then be reported to consumers at the point that they are making their decision about using the road. In either case, the goal is to have the road tolls be closely tied to the actual road conditions.

On dense central city streets, a toll on each street will not be an efficient solution to the congestion problem. The costs of administering the system will overwhelm the benefits from pricing congestion. A feasible alternative is to use cordon pricing, such as the system pioneered by Singapore. In this system, drivers need to pay to use roads within a given geographic area. As such, on a monthly (or weekly or daily basis) drivers will need to purchase the right to use the roads. This can be administered through a fairly primitive sticker technology where individuals who have paid will have to display a visible sticker. In principle, this can be replaced by transponder technology, which alerts the police if drivers have not paid for their driving privileges. This system has the advantage of taxing drivers who want to drive in crowded central city areas. However, it has a few disadvantages. First, it will often create congestion right on the edge of the cordon. For example, since Singapore had a hefty fee associated with driving within the cordon (roughly one dollar daily) and no fee outside the cordon, congestion on the edge of the cordon get quite high.

A natural response to this problem is to have several geographic zones, which allow the driving fees to drop continuously instead of having one big drop. A continuous drop means that the gap between one zone and another will be smaller and want cause huge crowding on the edge of the cordoned area. Of course, it is crucial that there be real price differences, between the areas because the whole point is to induce drivers to substitute out of the high congestion area. A second disadvantage with cordon pricing is that it doesn't differentiate by time of day. Driving at 2 a.m. is just as expensive as driving at 9:30 a.m. In principle, this can be soled by allowing individuals to drive for free very late at night. An even more complicated system allows an array of permits that get cheaper as the hours associated with the permit become more restricted.

While economists have been unreservedly enthusiastic about pricing tolls for decades, congestion tolls have only been rarely implemented. This lack of implementation is associated with the general political unpopularity of these tolls. In part this unpopularity comes from the fact that tolls are yet another tax. Since implementing congestion tolls is something of an uphill battle, it seems important to figure out ways to make tolls more political palatable. Here are three suggestions. First, give people some initial free credits on the tolls. These credits can either be used to pay for the tolls or to reduce an individual's taxes. This basically just gives the cash from the tolls back to consumers, but still gets each individual consumer to face the cost of driving since by driving during peak hours, they lose the opportunity to use the credits, which can be used to pay taxes. Second, tie toll revenue to a politically popular form of spending, perhaps new roads or schools. Tolls can be made less unpopular if it seen that the income is going on something sensible. Finally, use highway tolls primarily on new roads. People generally respond with more hostility if an existing right is modified (i.e. by charging a toll on a road that used to be free) than if a new right is given with qualifications (i.e. if a new road is built, but that road requires a toll). While their will still probably be protests about tolls on new roads, they will be much less severe than protests on existing roads.

As a final point on road tolls, it is worthwhile emphasising the winners and losers from these tolls. Many people will gain from these tolls through quicker commute times. Poorer drivers may lose a little if they cannot substitute into lower commute tax times of commuting. However, the truly poor will really benefit. As they use public transportation (buses) not cars, the road taxes that they pay will be much less (per capita) than drivers. These poorer commuters will get all of the benefits of faster commute times with a much lower commute tax per person. As a result, these commute taxes will generally be progressive and they will help the poorest commuters.

Given the political costs of tolls, it is worthwhile emphasising why other approaches to limiting congestion are so inefficient. First, the pure supply approach of building new roads is basically doomed to failure. New roads are surely necessary, but the basic market failure will not go away with new roads. Moreover, without time-based tolls drivers will inefficiently use any new roads at peak times instead of spreading their usage over the day. Second, quantity controls on driving inevitably involve large efficiency losses. A classic example of such quantity controls is license-based driving restrictions. According to these policies, certain cars must not be used on certain days.

There are certain obvious inefficiencies of this system. For example, it is clearly inefficient to have a sizeable amount of the capital stock of a city (i.e. its cars) lie idle 20% of the time. Moreover, these restrictions compel richer drivers to purchase extra cars so that they can use their cars every day. This does impose a cost on drivers, but this cost is pure social waste. Whereas a toll turns produces income that can be used for any other social purpose, spending on buying extra cars is just wasteful. These quantity restrictions also create less obvious social waste that can be avoided with tolls.

The beauty of road taxes is that they do not prevent people who really need to drive from using the roads. They just require that these people pay for the social costs of their actions. However, quantity controls require everyone to stop driving for some period of time. If those people really need to use the road during that time period, the law makes no allowance. The law doesn't allow the people who value the roads most to use them, instead it inefficiently blocks everyone. This is why economists almost universally prefer taxes to command and control restrictions.

14.4 Public-Private Co-operation and Partnerships

The creation of a public-private partnership offered a lot of advantages for financing the project as well as fostering operating processes. Planning stages were shorter than in an administrative way, flexibility in reacting on new developments was ensured. At the same time, it ensured the compliance with the ecological concepts.

14.5 Local Public and Private Parties in Land Use and Transport Planning

Vertical co-ordination does not leave much space for regional initiatives and is often seen as restrictive. Local governments ask for more and more flexibility and decentralisation. The local public and private parties should be prime actors, making a comprehensive regional land use and transportation plan without to much interference of national concepts, which are not adapted to their specific regional needs. The different parties should come to an agreement in a contract, which should contain measurable standards to achieve. The national government, which is one of the parties, could monitor the process using these agreed standards and adapt their financial support on the effectiveness of the local policy.

14.6 Project Financial Responsibility should be Borne by all Partners

Complex urban planning projects require a true political desire, which is defined in urban development documents and results in preserving land sites for the future even if it is a long term plan. They also need an appropriately set up structure to control all development work, through a public-private partnership. The financial risks of the project must be borne by all partners. In particular, banks, when present, must not limit themselves to being part of the structure simply to provide funds to operators and property developers who will be responsible for construction work.

14.7 Possibilities for Development of Sustainable Transport and Land Use Structures

It is important to identify measures in order to reach the development of sustainable transport and land use structures. These steps should not be isolated from each other. These measures can be divided in the field of spatial organisation and the field of transportation. To give an overview about measures in different spatial levels, case studies show projects, which were implemented in the national, regional, local, or district level.

Measures in the field of social integration, citizens communication and participation are useful to support these measures and are also taken into account. Spatial structures reducing distances between urban functions can be considered as an important precondition for decreasing traffic, especially traffic caused by motorised modes.

The measures can be subordinated under the following statement: a revision of spatial organisation or the development of spatial organisations in areas with future development that fit into traffic-reducing concepts. Vice versa, land use planning must recognise existing transport structures: for instance the development of a new area intending to promote less car dependency but with easy connection to a motorway may not be successful. The following measures could fit into an integrated concept: (a) Assign functions (housing, working, leisure, education, supply+services) on the urban/regional scale; (b) Development of mixed used structures; (c) De-central concentration; (d) Creation of dense, compact structures; (e) Protection of landscape and nature and reduction of land consumption by avoiding urban sprawl; (f) Redevelopment in existing structures; (g) Infill development on vacant land in existing structures; (h) Concentration of urban development around public transport stops; (i) Create concentrated development nodes which allow for public transport to be economically feasible; (j) Give attention to developing high quality public spaces to attract the public and create liveable cities; and (k) Measures of transport planning.

Land use planning should also consider existing transport structures. The following are examples for measures in the field of transportation planning that promote sustainability: (a) Promotion of public transport (allowing for quantitative and qualitative improvements in order to make public transport competitive with car use); (b) Improvement of accessibility of public transport; (c) Promotion of non-motorised transport (by making quantitative and qualitative improvements); (d) Promotion of transportation intermodal; (e) Influencing car usage (reducing traffic, traffic calming, parking management); and (f) Networking regional economy and production.

Referring to the sustainable triangle mentioned early, creating social equity is one important aim in urban sustainability. Social equity could be defined as giving all inhabitants of a city the opportunity to satisfy their demands and needs. Different social groups also have different demands on mobility. A consideration of these demands in an integrated transport and land use system could promote the accessibility of different social groups to cities facilities. People without cars should be given the opportunity to reach important locations by public transport, bike or walking. Conversely, the reduction of social inequity and social integration could influence spatial structures as well as the capacity of public transport. Reduction of social segregation could be related to public participation. Involvement of different social groups in decision-making helps to ensure that land use patterns meet the needs of these social groups. If mixed use facilities meet the demands of social groups living in a district than car mobility can be reduced. Some measures in this term are: (a) Increasing accessibility to transport networks; (b) Mix of private and public investors in building new districts; (c) Increasing accessibility to social, cultural life and working / education; and (d) Implementation of mix use to offer low distances for people without an own car.

Public participation as well as public awareness in order to ensure the acceptance of the above mentioned integrated planning goals must be considered as an important part of integrated transport and land use projects. Public information and participation can reduce social barriers which could hinder the implementation of planning tools promoting sustainability. Creating awareness among citizens about the disadvantages of car usage could, for example, promote the usage of public transport and non-motorised modes. The measures in this aspect are: (a) Create awareness to the public about mobility related problems; (b) Create awareness to the public for another mobility behaviour; (c) Influencing private housing preferences by awareness campaigns to promote higher population density; (d) Citizens' participation in planning and redeveloping of structures; (e) Citizens' participation in planning and implementation of new districts to promote local identity; (f) Informing citizens about the advantages of integrated measures; and (g) Information should be made available and participation by stakeholders, decision-makers and politicians involved in the project should be promoted.

The following groups have been identified as those who are targeted by implementation of best practice in the field of integrated transport and land use planning, who can use best practice and how they can profit from its use: (a) Planners - best practice as an example of a successful integrated planning approach. The successful implementation of a previous project can serve as a motivation. Such practices can help planners to choose the right planning measures for their particular project; (b) Citizens - must be informed that their quality of life will improve if they use best practice (this is very important especially with car use restrictions since such restrictions are not popular). Many measures towards integration could help reduce an existing social imbalance in cities. Certain social groups rely on public transport, bicycles, or travelling by foot due to objective and subjective dependency. Such groups profit by integration. Land use patterns (mixed use, density), public transport infrastructure investment and the promotion of non-motorised transport can facilitate access to urban functions by such social groups; (c) Decision-Makers - planning instruments which have been successful in the past are more likely to be accepted by policy makers; and (d) Investors - projects are more likely to be supported by investors if proven instruments or elements of best practice are used to reduce doubt by investors.

15. Lessons Learned

The construction of adequate public transport facilities is one of most problematic issues when implementing a housing policy. As large investments are necessarily required, it is safe to assume that considerable delays may be expected with negative pitfalls on the quality of service. These problems can be averted if (private) public transport companies were to be included at an earlier stage in the policy planning process.

Citizens' participation is pivotal for creating mixed-use areas. The shaping of traditional European cities leads to postulate that mixed-used areas are not the result of detailed planning – as they are more the outcome of spontaneous development. According to this assumption, self-organisation and citizens participation in developing mixed-used districts (not only in the planning and building stage) is one of the preconditions for creating a mixed-used area accepted by its inhabitants for a long time.

Shopping malls can compete with businesses located in well-integrated mixed-use city centres. Well-situated customers prefer shopping in malls, but also distant shoppers accept long travel distances, long travel time, as well as congestion in order to shop there. Campaigns informing that time is not gained when driving to non-integrated shopping centres in peak hours and public transport must be promoted to influence customers.

Location policies successfully regulate public and private investments and have strongly strengthened the vitality of the cities. Firstly, they can concentrate public investments in infrastructure and public transport within the urban areas. Secondly, they can start large urban renewal programs to upgrade the inner city areas around major transport node urban locations, and thirdly they help to attract private investments to the city. Especially the strong development of locations reachable both by public transport and car can induce a new economic impulse for the urban economy. To make a location policy successful, the implementation of other transport policies and land use policies are necessary. The location policy can only function well when included into a well-balanced policy package. Furthermore, the success of this policy depends on the availability of all the accessibility profiles. Refrain from creating new planning bodies.

While it is correct that the integration between land use and transportation planning is in its essence a regional task, it must be concluded that it is worthwhile using existing legislation as much as possible, before creating new institutional bodies to handle the planning. It is advisable to implement a procedure for creating a "zone of consistency between urban development and transportation" that should provide proper means of controlling the occupation of land as long as transportation infrastructures are not in place. This procedure imposes actual monitoring of the various urban development projects. Therefore, projecting a situation at a given time horizon is not enough, it is necessary to plan for the intermediate steps and control the mechanisms to be implemented in order to limit inconsistencies.

Lack of integration between transport and land-use can cause negative mobility effects such as increased share of motorised modes, as well as increased travel times and travel distances. A rigid and vertical planning hierarchy results in a series of strictly independent local plans organised in general, partial and special plans, action programs and detailed studies which effectively cause a disconnect and lack of co-ordination between authorities and citizens. Such extreme top-down articulation leads to a deadlock, which can only be overcome when a shift towards a more co-ordinated and participatory planning approach is decided. It is difficult for integrated transport to work in a semi-rural area due to poor public transport. Therefore, an alternative option to encourage sustainability is to encourage people to live in close proximity to their place of work.

The lack of building laws and regulations fosters growth of poorly integrated developments. Information is crucial during project development. In planning a light rail system, there is the need for detailed design and to keep the public informed during construction. It is crucial that great care is taken in ensuring that all parties know of construction work and the public kept up-to-date by media announcements. Light rail systems require specialist operational management. Concentrating too much on engineering, design and organisation and not considering the operation of the scheme can harm the project.

16. Transports Modalities

To evaluate the economic impacts of transports we can use some tools to estimate the effects of a transportation system intervention and then evaluate the screening of options. The commonly used tools are: Analysis of cost x benefit; Analysis of cost x effectiveness.

The automobile has a contradictory relation with the tertiary Sector. During decades the automobile industry in partnership with oil industries had dominated the economy. Eight of the ten bigger companies of the world.

The revolution of the computer science and the media if had inserted in this universe of being able, without the automobile industry lost its. Until the sixties it had the ideal of the motorised city. In the following years all the problems for this had been being eliminated: (a) Weakness of the railroad mass transport; Extinguishing of the trams of cities as Rio de Janeiro; (b) Bigger investments, each time, had been made in infrastructure for the individual transport: parking buildings and new facilities that have occupied the urban land; (c) Congestion; Deterioration of the collective transport; Subordination of the collective transport on the individual one. The cities sow its transport capacity be reduced.

With the urban growth and the concentration of the tertiary Sector in monocenters cities with structure, flows of bigger traffic each time if directed to the centres of the cities causing a jam of flows. From a certain moment the tertiary Sector observes diseconomy for price increasing of the workmanship and reduction of the demand due the inaccessibility generated for the congestion. We can see that more land occupation of automobiles, strangling the tertiary Sector, It's a cumulative and cyclical process that must be controlled.

The concept of "living without an own car" has encountered difficulties due to the overestimation of the demand for this style of life, and it offered the opponents arguments against the implementation of the new policies. Therefore, pilot projects must be based on a solid background in order to face off the barriers posed by detractors. At present the only method of enforcing car-free living is the prevention of residents' parking within the development, through the lack of parking spaces and through residents' good will.

In the decade of 60 and 70 urban decentralisation was not accepted for the tertiary Sector that prefers centrality. In the decade of 90 and 00 this tendency was reversibly because tax incentives for services, industries and others facilities that supply a degree of self-sufficiency of services experimented for some metropolitan zones such Barra da Tijuca, located in the west zone of Rio De Janeiro. The change of commercial building Esso Co. from city to Barra da Tijuca is an example. For many people cars are not a relevant means of getting to work. Instead, they use large public buses, subways and smaller private buses, or jitneys.

The subsidisation of public transportation is often vaunted as an alternative means of fighting congestion. While public transportation is quite important, economists tells us that subsidising buses is much less efficient that taxes cars are a means of fighting congestion. However, while buses shouldn't be subsidised, and indeed in principle buses should also be taxed for the congestion that they create, they are an extremely important part of the urban transportation landscape. They provide a very efficient means of moving poorer people to and from work. In particular, smaller buses, or jitneys, provide an unusually efficient means of getting poorer people to their jobs. As such, they should be recognised as an extremely valuable part of the urban transportation system.

Regulatory barriers should not prevent these jitneys from operating. Except for the principle of taxing congestion, there is no reason why free entry shouldn't be allowed and encouraged in the bus system. If this is allowed to go forward, there is no reason to doubt that Brazil will continue to have a healthy private bus system that delivers people to jobs. While buses are a efficient means of moving, subways are generally expensive in construction and operation. They are generally sold to the taxpayers with a variety of gimmicks, such as vastly over inflated rider ship projections. Serious economic estimates of the costs of subways tell us that these subways, for any level of rider ship are much less cost effective than buses.

A successful scheme is to have buses on dedicated bus lanes. Buses on these lanes can move as quickly as trains, but there is much more flexibility. Given the unpredictability of cities, it makes no sense to invest in expensive fixed infrastructure that can never cover its operating costs, let alone its construction costs.

Great investments on bikeways, and not just in the central city zones, seeking to form an attractive network, have increased the cycling share providing good cycling opportunities also for inhabitants living outside the downtown area. Such investments, suggest that cycling can be promoted as a mode of transport in both high and low density areas.

When promoting cycling, conflicts between pedestrians, car drivers and cyclists will inevitably arise but they can be solved with information campaigns, and by means of restrictions (dividing pedestrian and cycling areas, prohibition of cycling in pedestrian areas and vice versa). Interesting design and landscaping will encourage potential residents who are considering living in a car-free environment.

17. Conclusions

Land-use and transport policies are only successful with respect to criteria essential for sustainable urban transport (reduction of travel distances and travel time and reduction of share of car travel) if they make car travel less attractive (i.e. more expensive or slower)(Transland, 2000).

Land-use policies to increase urban density or mixed land use without accompanying measures to make car travel more expensive or slower have only little effect, as people will continue to make long trips to maximise opportunities within their travel-cost and travel time budgets. However, these policies are important in the long run as they provide the preconditions for a less car-dependent urban way of life in the future.

Transport policies making car travel less attractive (more expensive or slower) are very effective in achieving the goals of reduction of travel distance and share of car travel. However, they depend on a spatial organisation that is not too dispersed. In addition, highly diversified labour markets and different work places of workers in multiple-worker households set limits to an optimum co-ordination of work places and residences.

Large spatially not integrated retail and leisure facilities increase the distance travelled by car and the share of car travel. Land-use policies to prevent the development of such facilities are more effective than land-use policies aimed at promoting high-density, mixed-use development. Fears that land-use and transport policies designed to constrain the use of cars in city centres are detrimental to the economic viability of city centres have in no case been confirmed by reality (except in cases where at the same time massive retail developments at peripheral Greenfield locations have been approved).

Transport policies to improve the attractiveness of public transport have in general not led to a major reduction of car travel, attracted only little development at public transport stations, but contributed to further sub urbanisation of population. In summary, if land-use and transport policies are compared, transport policies are by far more direct and efficient in achieving sustainable urban transport. However, accompanying and supporting land-use policies are essential for in the long run creating less car-dependent cities. The leading objective of land-use and transport planning is to reduce the need for travel and to promote sustainable transport. Different policies were assigned to policy types: investment and services, planning, regulation, pricing and information, and informal policies.

Due to their interdependent effects policies of land-use and transport need to be combined to reach the sustainable objectives. This mainly refers to the relationship of investment and services and planning on the one hand and regulation, pricing and to a certain extent information on the other hand. Most policies relating to planning and investment, while necessary, are not adequate by themselves to reduce the need for travel and to reach sustainable transport. Their successful implementation is only possible if additional pricing and regulatory policies create the necessary frameworks.

Planning and investment policies are nevertheless the most important means to reduce the need for travel because they influence land-use, traffic infrastructure and travel behaviour. However, they often must be coupled with pricing and regulatory policies, which not only support the planning and investment policies but also promote a change in the settlement behaviour, a reduction of land-consumption and support an efficient use of the transport network. It can be concluded that all policies are important and they can be used in combination to lead to successful implementation.

The realisation of the policies can be restricted or prevented by different types of barriers, including resource barriers, social/political, legal and institutional barriers, as well as side effects. These barriers determine the feasibility and transferability of policies. It can be concluded that all policy types, except information policies, face several barriers, with planning and investment mainly being restricted by institutional barriers and pricing and regulatory policies mainly facing social barriers. Information policies, which are limited in their effects on reducing the need to travel, hardly face any barriers. The future of cities is intrinsically bound up with transportation technologies.

Cars have changed urban form and will continue to change urban form. However, unless the congestion problem is solved, cities will not hobbled with the extremely costly problem of long commute times. The congestion problem is a classic externality problem where drivers don't take into account the cost of their driving on others. The best solution to this problem is a traffic toll that is targeted at specific roads during specific time periods. Less direct taxes will be much less efficient at reducing congestion.

Quantity controls, such as license plate based restrictions on days of driving, are also quite inefficient. Public transportation will continue to play a role in urban transportation for the foreseeable future. However, all of the economic analyses of public transportation suggest that subways are extremely costly, inefficient means of solving the transportation problem.
There is nothing that a subway can do that can't be done better by a dedicated bus line. As such, it is crucial that Brazil not wastes money only in an expensive subway extensions but rather improve the bus infrastructure and operation instead.
To avoid the stimulating automobile use, a good and relatively cheapest option is to turn some shared road spaces into preferential and/or exclusively bus lane. This practice will convert mix bus/auto modal to bus priority modal capacity and will turn less attractive the use of automobile, in accordance with Transland studies. This solution only will be effective if the lane will be in the same direction of jam flow, otherwise the use of automobile will be increased and the problem will be greater.
The level of service of bus modal in preferred or exclusive lanes must be better than before. Time will be reduced by the speed increase naturally. There must be interventions in the end of exclusive lanes and intermodal terminals in such a way to turn the solution integrated and equilibrated. To avoid the interference of right turns, the access of buses could be made by a left door, utilising the central spaces of streets. Bikeways integrating this buses terminal could be a good solution for short and meddle distances of commute and also complain with ambient aspects.

18. References

Ben-Akiva, M.E. & Lerman, S. (1985). *Discrete Choice Analysis: Theory and Application to Travel Demand,* Cambridge, MA: MIT Press

Ben-Akiva, M.E. (1974). *Structure of passenger travel demand models,* Transportation Research Record 526, pp. 26-41

Ben-Akiva, M.E., Bowman, J.L. & Gopinath, D. (1996). *Travel demand model system for the information era,* Transportation 23, pp. 241-266

Developing the citizen's network (1998). *Why good local and regional passenger transport is important, and how the European Commission is helping to bring it about,* COM (98) 431 final

ESTEEM Consortium (1998). ESTEEM, Final Report, Roma: ISIS

Green Book (2004). *A Policy on Geometric Design of Highways and Streets,* AASHTO, 5th Edition, ISBN Number: 1-56051-263-6

PDTU (2001). *Plano Diretor de Transportes Urbanos,* Governo do Estado Rio de Janeiro, Secretaria de Estado de Transportes, Companhia Estadual de Engenharia de Transportes e Logística, Central

Transland UR-98-RS-3055 (2000). *Integration of Transport and Land Use Planning,* Paulley, N. & Pedler, A., Transportation Research Laboratory – TRL, London

Williams, K. (2005). *Spatial Planning, Urban Form And Sustainable Transport (Urban Planning and Environment),* Ashgate Publishing, ISBN-10: 0754642518 ISBN-13: 978-0754642510, United States

3

Efficient Solutions for Urban Mobility - Policies, Strategies and Measures

Alvaro Seco, Ana Bastos Silva
University of Coimbra
aseco@dec.uc.pt, abastos@dec.uc.pt
Portugal

1. Introduction: Formulation Processes of Mobility Policies

Over the past few decades, particularly in urban areas, mobility needs have significantly grown and changed as a result of the normal social and economic development. The mobility is nowadays a very diverse and complex reality, in reason of the tendency for a more disperse residential occupation and for a more decentralized location of most commercial and service activities, as well as of different population mobility habits resulting from their increased wealth. As a consequence urban mobility has been ever more dependent on the private car and, in many cases, by the existence of inefficient and costly public transport systems, with obvious negative impacts at the environmental, social and economic levels for the society as a all.

It is also relevant to refer that in some European Union (EU) countries transports use up to 30% of the energy used by the different human activity sectors and is responsible for 25-30% of the total of greenhouse gases (EEA, 2000; Civitas, 2006), with the car being responsible for as much as 50% of the emissions produced by passenger transport systems.

It is also important to notice the negative impacts that transport systems can, and often have, over several quality of life aspects. In many cases these systems invade many of the cities public spaces, which are otherwise used in many other activities such as leisure.

This situation has led to an increased emphasis being placed in the development of transport strategies and solutions within the Sustainable Development Global Agenda (Commission of the European Communities, 2006). The EU Green Paper over Urban Environment, the EU Treaty, the successive EU environment and transport action programs, the Rio de Janeiro UN Conference on Environment and Development or the different UN conferences culminating with HABITAT II, constitute some of the initiatives witch have been raising the sustainability issue and, in this context, have been discussing the future of urban mobility.

An urban strategic planning process, taking into consideration the urban area fundamental characteristics and its population needs, is thus an essential framework for the identification of adequate sustainable transport policies.

These planning processes can vary significantly but generally it can be said that they are evermore inter-disciplinary and focused mainly on two different but complementary areas.

One focus is on the identification of packages of measures directed at achieving an effective modal shift towards the most sustainable ones and the other directed at achieving a reduction of the need for transport.

In this context the formulation of a mobility policy applicable to complex urban environments and which can serve as a supporting basis for subsequent planning, implementation and management of transport schemes, is a complex process where many technical and political questions and decisions interact and which involve a very significant number of stakeholders. It is, however, possible to define a number of basic methodological principles, as well as some typical system intervention strategies and measures, which can work as a framework to this process.

The first step of the process (see Figure 1) deals with the identification of the existing problems and of the basic strategic objectives which are to be achieved with the implementation of the new mobility policy. At the same time the definition of a set of performance evaluation criteria, applicable both during the initial diagnosis phase and during the final evaluation and monitoring periods, is essential.

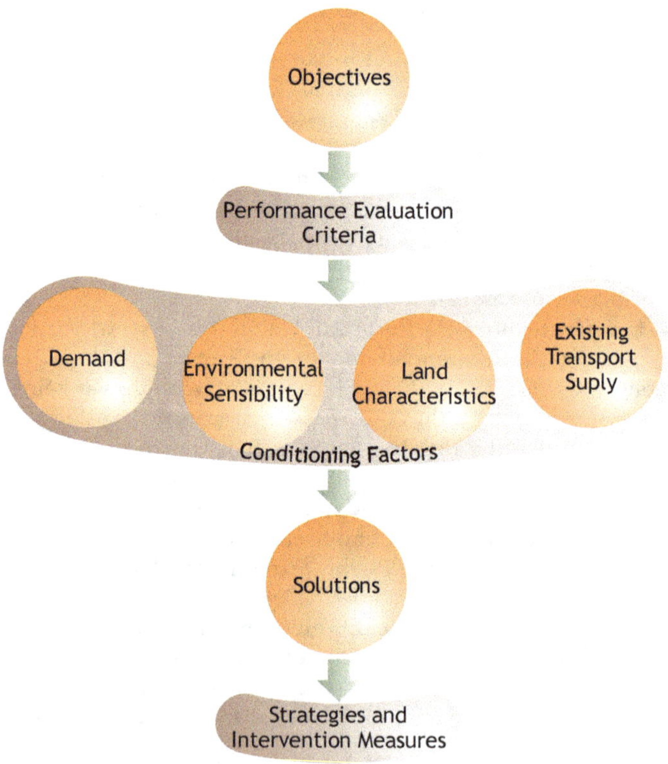

Fig. 1. The process to formulate a new transport policy

The second step focus on the identification and characterization of all the factors which, some how, limit the universe of ways in which the transport system can be structured.

A basic conditioning factor is in itself the way society view and value the concept of quality of life, namely in regard to the natural and historic heritage, and how it views environmental sustainability problems which result from the way society in general and the transport system in particular is organized.

Other important conditioning factors are, of course, the potential, weaknesses, and flexibility to change the existing transport systems have. Similarly important to be considered are the existing levels of transport demand and supply, and their predictable evolution in the foreseeable future. In fact the demand patterns, which result from the existing economic and social practices, as well as the specific characteristics of the existing transport supply systems, create a significant inertia and restriction to the eventual selection of new organizing and operational transport solutions.

In a similar way the specific natural and built land characteristics will be of paramount importance to the selection of efficient solutions and thus will need to be particularly well known and understood.

The understanding of the ways in which all these different conditioning factors interact enables the identification of the most efficient transport system organizing solutions, which will tend to be drawn and adapted from a number of "typical" ones.

In the present text reference is made to generally adequate organizing solutions applicable to different urban environments, namely those who were designated as "Historical Areas", "Traditional City Centers", "Modern, Medium-High Density Developments" and "Suburban, Low Density Developments". The "scale" of the problem is a topic which also needs to be taken into consideration in any process of this kind and, thus, will be briefly analyzed.

Having identified the adequate transport policy to be adopted, it will then be necessary to select a coherent set of basic intervention strategies and measures capable of guarantying its adequate implementation.

In the current text the different strategies and measures which are generally applicable are presented in a structured way, with reference being made not only to their potential but also to their applicability conditions.

In the final part of the text, a number of real life benchmark case studies are presented, in order to better demonstrate the potential that exists to implement efficient and sustainable transport policies.

2. Transport Policies' Objectives

Although the specific solutions adequate for each urban space will decisively depend on their specific mobility problems and of its own population and their representatives perspectives, it is however possible to identify a set of strategic objectives which are relatively consensual and that can work as basic references in any urban mobility policy defining process. Three main strategic objectives which are increasingly consensual can be identified:

- To contribute to the improvement of the populations quality of life by guarantying the provision of good and equitable mobility conditions for all;
- To contribute to the economic development, through the provision of good accessibility by people and goods to the different spaces of the territory;

- To optimize the global efficiency of the transport system both in operational and energy and environmental terms.

This means that the transport system must also assume a social mission by guarantying mobility to every one, including people with special mobility needs, as well as adequate accessibility to all urban areas, including the ones with scarce occupancy, even when that is not economically sustainable. It is also important to notice that the efficiency of the system is based on the optimization of its operation, especially at the speed, reliability and safety of the offered services, as well as by minimizing of the financial effort associated with the implementation, operation and maintenance of the system. The minimization of the negative impacts that the functioning of the system inevitably have over the natural and urban environment, is also gaining significant importance, as is the option towards the adoption of more energy efficient, less dependent on hydrocarbon fuels solutions.

The set of general objectives identified above imply that any transport problem will inevitably be a multiple objective one. Furthermore the increasing importance of aspects like safety, minimization of energy dependency and, specially, minimization of environmental impacts has created the need for a more systematic identification, quantification and taking into consideration of all the costs and benefits involved in the operation of a transport system, namely through a process of internalization of the transport externalities.

3. Transport Policies' Conditioning Factors

As referred above the development of efficient and sustainable transport policies must take into consideration a number of technical and sociological conditioning factors namely: demand basic characteristics; type of land use and natural characteristics of the territory; existing and implementable transport modes and services; residents and stakeholders' environmental awareness.

The demand patterns are a consequence of the location and relative importance of the different land use types, but are also the result of their scale and concentration levels.

The scale and density of the land usage tend to decisively influence the eventual existence and typology of the public transport systems. Banister (2007), for example, states that some empirical studies suggest that some level of sustainability for quality public transport systems can only be achieved in cities with at least 25,000 (preferably 50,000) inhabitants, with occupancy densities of at least around 40 hab/ha.

Also the population social-economic basic characteristics can significantly influence the potential applicability of different transport solutions. Households' income is a major factor affecting private car ownership levels which tend to significantly influence private car mode share. Also the age distribution of the population might be an important factor, for example to determine the potential importance of school trips.

Land characteristics can also significantly condition the choice of transport supply solutions. The existence of a significantly consolidated urban area immediately imposes significant difficulties or costs to solutions which might imply the implementation of new infrastructure components. On the other hand, the existence of important historical or natural spaces will tend to force the adoption of less intrusive, more environmentally friendly solutions. Different natural barriers such as rivers or other significant physical terrain features can significantly influence transport networks shapes and characteristics.

Also the type, potential and performance of existing transport infrastructures is an obvious and decisive conditioning factor since seldom a transport problem occurs in an area with very little existing infrastructures and systems, leading to the option of optimizing those instead of introducing new ones, being very attractive. The basic characteristics of the different readily implementable new systems is also an important factor to be taken into consideration since, in most cases, the development of an entirely new system is not an option. The thorough knowledge of all applicable systems and services is thus of paramount importance when developing a structured transport policy and selecting the most adequate intervention strategies and measures. The basic characteristics, but also the potential and applicability conditions of some of the most common and interesting modes and services are presented below.

Finally, it is also important to make a reference to the conditioning factors associated with the different stakeholders' sensitivity in relation to different aspects connected to the concepts of quality of life in general and environmental quality and sustainability in particular. The fact that in many countries and societies the possession of a private car gives social status, while the usage of, for example a bicycle, is a sign of low social status, introduces significant difficulties to the consideration of more environmentally friendly solutions. Also the existence of a more or less intense environmental consciousness by the different stakeholders but, especially, by the potential users, can significantly affect the success prospects of policies where this aspect of the problem is important.

At this level the quantification of the externalities associated with the operation of the different modes and services is very important, since one of the potentially more important strategies to guaranty the successful implementation of more sustainable policies is the internalization by each system or service of their intrinsic externalities. This aspect of the problem if further analyzed below.

4. Transport Systems: Characteristics, Potential and Applicability

At the present many transport systems and services can be considered for application, ranging from the more traditional pedestrian, bicycle, private motorized vehicles, or road and rail based collective modes, to the more innovative ones such as car sharing or car pooling. There are also other more or less specialized solutions applicable to special problems such as the mechanical elevatory or maritime systems or, finally, those which involve the integrated use of more than one mode, such as Park-and-Ride or Bike-and-Ride.

Each of these systems and sub-systems presents specific intrinsic characteristics, both at their operational and performance potential levels, which decisively influence their applicability to the resolution of the different mobility problems which might exist in an urban area. Table 1 presents a brief characterization of some of the more relevant and common modes (see for example Vuchic, 2007).

The pedestrian and cyclist modes offer an excellent timing availability although, sometimes, this can be limited by adverse weather conditions. On the other hand, these modes' spatial range, particularly that of the pedestrian mode, are somehow limited, not only due to the limited distances which can be covered in view of their limited operational speeds, but also due to their difficulty in dealing with adverse orography.

	Pedestrian	Bicycle	Auto Taxi	P&R B&R	D. Ride BUS	Light Rail	Metro Rail
Service potential							
• Operational Speed(km/h)	3/5	10/20	20/70		12/25	20/45	25/70
• Coverage Range (km)	1	5/10	≈40		≈20	≈40	≈40
• Capacity (Pas./h)x10^3	4,5 (/meter)	2,0 (/lane 1,2m)	1,4/2,2		2,4/8	6/20	10/40
• Productive Cap. (Pas.xKm/h^2) x10^3			10/120		20/90	120/600	400/2000
• Availability/Frequency	Very Good	Very Good	Very Good	Good			
• Spacial Availability	Good	Good +	Very Good	Good +			
• Comfort			Very Good	Good +	Good	Good +	Good +
Implementation/operation							
• Adaptability/phasing possibility /Level of Investment	Very Good	Very Good	Very Good	Good	Good		
• Energy Efficiency/ environment/intrusion	Very Good	Very Good		Good	Good	Very Good	Very Good
Preferential application							
• Link type	Short distance	Level terrain	Low -Low density	Low - Med./high density	Med. - Med./high density	Med./high- Med./high density	High- High density

Table 1. Potential, implementation and operation conditions of some transport modes

In operational terms both the pedestrian and cyclist modes present interesting potential due to their implementation and operational easiness, since they can be implemented progressively and do not need sophisticated management and control systems.

They also present the highest energy and environmental efficiency levels and the smallest urban intrusion impacts.

All these characteristics imply that their competitiveness is highest in dense urban developments where trips will tend not to be very long. Furthermore, these modes are especially interesting in the implementation of aggressively sustainable policies.

The private car is characterized by its unbeatable timing and spatial flexibility and by its intrinsic comfort. In fact, no other mode can match the freedom that the car can offer to go almost anywhere at any time in completely private conditions and in complete comfort offered by, amongst others, their air conditioning and audiovisual systems.

However, at the present, it is also the most inefficient mode at not only the energy and environment levels, extremely important aspects in terms of achieving a sustainable mobility, but also at the intrusion of urban spaces level and in terms of efficiency of use of the road networks' capacity. If it is very likely that in time the first two aspects might be less conditioning factors due to the expected development of more efficient, less polluting, propulsion technologies, on the contrary it is very likely that the other two will maintain their importance.

An overall evaluation of all these characteristics leads to the conclusion that this mode of transport is especially competitive in not very environmentally sensitive urban contexts and where there is limited concentration of trips within the territory and, particularly, when there are not only good road connections but particularly good parking conditions at the destination locations.

It is also worth nothing the existence of a number of related sub-modes as are the Taxi, or the Rent-a-Car, the Car-Sharing or Car-Pooling systems or, in a slightly different perspective, the Motorbike, which present slightly different characteristics and applicability potentials enabling the coverage of specific market niches.

In contrast with the private car, road based collective transport systems present the possibility of offering significant higher transport capacities and lower urban and environmental impacts while using similar infrastructures' space. On the other hand they offer less scheduling and geographical coverage.

This leads to them being considered potentially more efficient and sustainable if they are applied in medium-high demand concentration urban spaces and to serve trips which are simple, for example single destination ones, and repetitive in geographical and timing terms.

A number of different sub-systems and services are also present within this mode, as are the Dial-a-Ride and the Metro Bus (where there is a systematic use of segregated lanes), or even, although less distinctive, the services resulting from the use of different types of vehicles such as Mini Buses or Articulated ones. All these solutions enable a significant enlargement of the applicability field of this type of systems.

Rail based systems such as trams, tram-trains, metro or regional trains, all present some characteristics similar to those of the road based collective systems. However they present a potential for much higher capacity levels and for offering higher operational speeds, and have the potential for offering the highest performance in energy consumption and environmental terms. On the other hand, they need a special infrastructure, generally segregated from the other urban spaces (not determinant but very useful in the cases of Trams and, particularly, Tram-Trains), which is much more demanding in terms of initial financial and time investment and require more sophisticated management and control systems. Relative to the road based collective systems they also present less adaptability to significant shifts in demand patterns, thus requiring more sophisticated planning systems.

All these characteristics make these systems particularly competitive when serving links and networks which connect high occupation density areas (>50 hab/ha) where very high number of trips are generated.

In terms of range tram and metro based systems are particularly suited to serve short-medium distance, urban trips with high frequency of stops, while tram-train and regional train systems serve more medium-long distance suburban and regional trips, with less stops and higher commercial and operational speeds.

Within the more urban solutions the decisive difference is that while the complete segregation of the metro enables higher commercial speeds and capacity, the tram solutions enable a closer, less expensive service in smaller urban areas.

The more recent Tram-Train solutions use vehicles with special technical specifications which enable them to function basically like Trams within the city centers and as regional trains across the suburban environments.

Within the multimodal solutions it is worth a special reference to the Park&Ride/Metro/Tram or Bike&Ride/Metro/Train ones, since they are amongst the more common and with more potential.

In these solutions an individual mode is intertwined with a collective mode of transport at a certain interface where, generally, there exists a long term parking area and a collective transport station.
In some cases, when the bike mode is involved, instead of parking the bike near the station, it is carried in the collective mode of transport, transforming the system in a Bike&Ride/Metro/Train&Bike one.
The combined usage of two very different transport modes enables the implementation of services with special characteristics where, basically, on one hand one takes advantage of the greater timing and spatial flexibility of the individual modes to serve the part of the trips which takes place in low-medium density demand areas, and of the higher transport capacity and efficiency in using urban space or higher range provided by the collective modes to serve the high urban occupancy areas.
These types of solutions are, thus, particularly competitive when connecting low density, suburban areas with high density, urban ones and, in particular, to serve more stable, repetitive home-to-work and home-to-school trips.

5. The Problem of Transport Externalities

Associated with transport systems operations one can identify different costs which can be classified either as internal or external. The internal ones are those which are directly beard by the users, while the others, more of a social type, are generally supported by the society as an all either at a local or at a global scale.
The justification for the adoption of a strategy of internalization of all the externalities is basically one of adopting a user-payer logic or, perhaps more appropriately, beneficiary-payer, meaning that who benefits from the provided service should bear all the associated costs, so that everyone is encouraged to adopt travel behaviors taking in consideration all the associated costs, including the social ones.
This strategy is off course essential in order to make the competition between all modes of transport "fairer" and in order to be able to create competitive conditions for the more environmentally friendly modes which are essential in the creation of a more sustainable mobility system.
However, external cost quantification processes present significant technical and political difficulties. At the technical level a basic difficulty is the choice of the most adequate type of analysis. One alternative is based on the quantification of induced costs, "damaged costs methodologies", which is normally carried out using declared preferences techniques, with which a quantification is made of the amount of money users are willing to pay to avoid the damage or willing to receive to accept it (WTP or WTA). The alternative methodologies, designated "avoidance costs methodologies", are based on the quantification of the costs associated with avoiding the occurrence of the external costs expected to occur if nothing is done to avoid them.
It should be noticed that this choice of methodology has significant implications on the obtained results (see Table 2) since the "avoidance costs methodologies" tend to produce much higher estimates than those based on "damage costs" (Austroads, 2003).
Still at the technical level there exist significant difficulties, on one hand, to identify all involved costs and, on the other hand, to accurately predict the future evolution not only of the phenomena that are responsible for the costs, but also of the exact values of these ones.

		Average cost per Passenger PT1995 (Euros/1000pkm)		Average cost per Passenger EU1995 (Euros/1000pkm)		Average cost per Passenger EU2000 (Euros/1000pkm)		Average cost per Passenger EU2010 (Euros/1000pkm)	
		Vehicle	Bus	Vehicle	Bus	Vehicle	Bus	Vehicle	Bus
Road Accidents		35,0	2,7	35,7	3,1	30,9	2,4	41,9	3,6
Noise		2,0	8,0	3,7	1,3	5,2	1,3	6,0	1,6
Atmosferic polution		8,0	6,1	17,3	19,6	12,7	20,7	13,9	19,4
Climatic changes	High	10,0	4,5	15,9	8,9	17,6	8,3	17,8	12,0
	Low	± 1.4	± 0.7	± 2.3	± 1.3	2,5	1,2		
Nature and Landscape		1,0	0,2	2,3	0,8	2,9	0,7	3,0	1,0
Urban impact		0,0	0,1	1,3	0,5	1,6	0,4	1,6	0,5
Before and after processes		5,0	1,8	3,6	4,3	5,2	3,9	8,3	5,1
Traffic jams		1,2	1,3	5,8	3,1			± 3.4	± 2.9
Total without traffic jams and Climate changes-High		61,0	16,0	87,0	38,5	76,0	37,7	92,7	43,1

Notes - References

Ref: INFRAS/IWW (2000/2004)

Values adapted from the study

Notes – Climatic changes scenarios

Method "Damage Cost - Willingness to Pay" less restrictive than "Control Cost - Avoidance Cost"

Used method "Control Cost" - scenario "High" - 140 €/ton CO2 (50% reduction of CO2 from 1997 to 2030)

Used method "Control Cost" - scenario "Low" - 20 €/ton CO2 (8% reduction of CO2, 1990 to 2010 - Kyoto)

Table 2. Transport external costs according to Infras and IWW (2004)

The political level, on the other hand, leads to a strong subjectivity in this type of process since, for example, when one uses "avoidance costs", the final results depend strongly on the adopted targets relatively to the reduction of the phenomena which produce the costs.
For example, according to Infras and IWW (Infras/IWW, 2000/2004) in order to achieve the Kyoto targets, which aim for a CO2 reduction between 1990 and 2010 of 5% at the world level and of 8% at the European Union level, it implies the assumption of a 20 Euros/tonCO2. If, on the other hand, the adopted objectives are those proposed by the UNFCC, "UN Framework on Climate Change", which aim for a 50% reduction at the World level and an 80% reduction at the OECD level in 2030 in relation to 1997, then it results in an estimated cost of around 140 Euros/tonCO2.
Also, when a methodology based on the WTP or WTA principles is used, the estimated costs will depend significantly on several factors. For example, the income levels of the population being questioned will probably be a relevant factor since, normally, the population willingness to assume costs will tend to be greater the greater their incomes are.

6. Sustainable Mobility Policies: Reference Solutions

6.1 Efficient versus Optimal Solutions

In multiple objectives, complex, problems it is usually impossible to identify optimal solutions, since conflicting objectives tend to coexist and it is not always possible to refer them all to the same measuring unit. In the transport field this tends to result in complex performance evaluation processes to access alternative transport systems organizing solutions. For example, objectives like minimization of the investment effort or performance optimization, on one hand, and minimization of the environmental impact and energy consumption reduction on the other hand, are conflicting and not easily reduced to a single monetary unit.

This leads to the search for "efficient" rather than "optimal" solutions where one identifies which, amongst the solutions which are the best regarding one or more partial objectives, are also the most efficient regarding all the remaining objectives. This concept can be visualized on the example represented on Figure 2 where one can say that the solution represented by the continuous red line is "better" and thus more efficient than the one represented by the dotted red line, but one cannot necessarily conclude the same when comparing it with the solution represented by the continuous blue line.

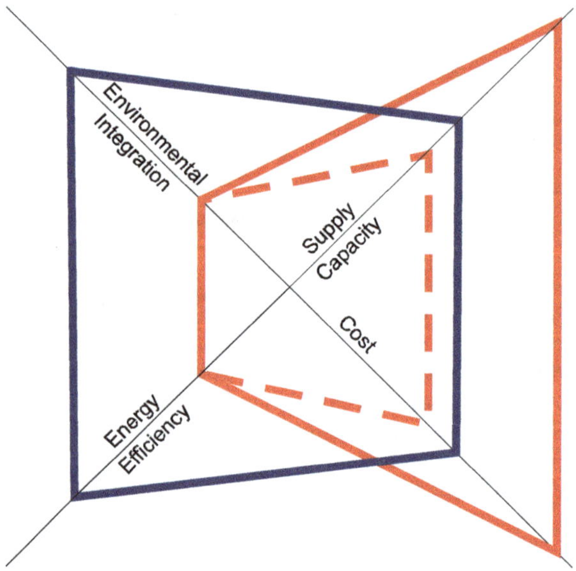

Fig. 2. Efficient versus Optimal Solutions

In complex transport problems this means that in many cases it is not possible to identify truly optimal solutions. However, from the identification of integrated solutions which are very efficient in regard to a significant number of objectives and which fulfil at least minimum requirements in relation to all the other objectives, it is normally possible to identify a reduced number of alternative "efficient" solutions. In particular for a number of characteristic transport problems occurring in certain representative urban environments, using adequate benchmarking it is possible to identify "efficient" integrated policies, intervention strategies and measures, which can confidently be applied.

In the current section four different and representative urban environments are analyzed and, for each of them, a set of basic options and solutions is identified as capable of creating efficient accessibility and internal mobility conditions. The urban environments object of analysis were "Historical Areas", "Traditional City Centers", "Modern, Medium-High Density, Developments" and "Suburban, Low Density Developments".

It is however worth noticing that the variability of the characteristics and potential presented by the different transport modes and services which can be applied is such that, although the types of policies which are considered adequate for a certain type of environment are reasonably similar, on the contrary the specific solutions to be applied can vary significantly and are particularly dependent on the "scale" of the problem at hand. In fact, for example, the mobility problems and consequent applicable solutions related to small urban developments with 15/20.000 inhabitants are necessarily different from those with 100/150.000 inhabitants ones and, obviously, even more from those of big metropolitan areas.

6.2 Efficient and Sustainable Solutions for Historical Areas

Any solution to be selected for application in an Historical Area must have as basic reference the nobility and intrinsic quality of the urban space. On the other hand, from a transport infrastructure point of view the main reference tends to be their extreme irregularity and limited potential (see examples in figures 3 and 4).

Fig. 3 and 4. Details of the historic areas of Coimbra and Viseu in Portugal

From these two basic factors it results that the existing and potential motorized capacity is very limited since, on one hand, there tends to be inadmissible any significant change in the infrastructure and, on the other hand, even when the potential operational capacity is significant, the real operational usable capacity will tend to be quite moderate due to the application to the road network of the concept of "environmentally sustainable capacity".

From this it results that both the accessibility to and mobility within historical areas must essentially be guaranteed by public transport, by special mechanized modes whenever appropriate, and by other environmentally friendly modes such as pedestrian and bike ones.

In fact, an efficient and sustainable usage of the very limited available road capacity implies that the essential of the access to these areas must be guaranteed by public transport, if possible completely ecological, with private motorized vehicles' usage being reserved and even so in a restricted way to priority users (residents, load and unload activities, priority and emergency vehicles, and people with special disabilities).

Special attention must of course be given to an adequate interconnection between internal transport modes and those that serve the surrounding areas. A relevant example is the possible interconnection between surrounding car parking areas and the internal pedestrian network complemented where relevant by mechanical elevatory systems or other internal public transport services. The same modes and services will of course constitute the backbone of the internal mobility system.

6.3 Efficient and Sustainable Solutions for Traditional City Centers

Most traditional city centers are characterized by the significant importance of commerce and services, which involve significant numbers of trips towards and from these areas with significant concentration during rush hours. At the transport infrastructure level it is common to exist road networks with limited capacity in relation to the potential demand, due to the fact that, in many cases, they were designed and built at a time when the private car did not have the dominant role it now tends to have. At the same time, because normally these are consolidated areas, there is very limited space to significantly expand the transport infrastructure unless underground solutions are assumed (see figures 5 and 6).

Fig. 5 and 6. Areas of the City Center of Coimbra in Portugal

Besides, having in consideration the desirable existence of quality public spaces, for which it is always negative the existence of high levels of motorized traffic, it will often be justified also to apply the concept of environmentally acceptable road capacities, although with significantly higher acceptable levels than those normally assumed in historical areas.

As a result of all these factors it is normally virtually impossible to serve most of the home to work movements by private car with any quality and without major impacts over the environment and the city quality of life. Within this context it is clear that the access to this type of urban areas, particularly by home to work type of movements, must be served by traditional public transport or by P&R services, with the exact mix of allocated services mainly dependent on the geographical pattern and intensity of the corresponding flows of each specific situation. On the other hand accessibility by commerce and services users, as well as by residents, should usually be served by all the available modes and services in "loyal" competition. To enable this it is necessary that the users bear all the costs for which each mode or service is responsible including those relating with "invasion of the urban space" and with the environment. In what it concerns residents, within coherent strategies against the desertification of the city centers, in many cases it might be advisable to implement positive discrimination solutions such as priority given in the access to public car parking.

The internal trips should be mainly served by the more environmentally friendly modes, particularly pedestrian, for which it is essential that this mode is provided with dense, comfortable and safe infrastructure networks directly connecting all the important trip generation equipments.

Finally, in what concerns the best use of the road networks maximum usable capacity, all efforts should be made to eliminate through road traffic since it does not bring any value to these areas. At the same time, it will normally be justifiable to manage the existing road network capacity giving priority to the most efficient modes (collective and or more environmentally friendly), namely using a logic of maximization of the number of people rather than the number of vehicles susceptible of being served.

6.4 Efficient and Sustainable Solutions for Modern, Medium-High Density Urban Areas

The more recent, medium-high density, urban areas in many cases present residential occupancy levels in the order of 60/100 hab/ha and, in most cases, have already been designed, although sometimes inadequately, with the road networks and accesses needed for a more car oriented way of life (see examples in figures 7 and 8). In these cases it is normally acceptable to serve most accessibility needs using all the modes available, providing that all the corresponding costs, direct and indirect, are internalized and supported by the respective users.

Fig. 7 and 8. Examples of Medium Density Neighborhoods in Coimbra, Portugal

In order to give competitive conditions to the public transport and bicycle modes it is essential that inside these areas adequate infrastructures are created along the full length of the trips, so that real door to door services can be provided. Public transport modes need not only comfortable and well localized stops but also a coherent interconnection with the pedestrian and cycling networks. These environmentally friendly modes should also be the main support for the internal trips for which it is essential that there exist dense, comfortable and safe networks, where one of the main aspects to be taken care of is the adequate management of the conflicts between these and the road networks, where the control of the vehicles' speeds tend to be very important.

6.5 Efficient and Sustainable Solutions for Suburban, Low Density Developments

Suburban residential areas are usually characterized by land uses with densities in the order of 10/25 hab/ha resulting from a mix of fundamentally rural areas and small urban agglomerates (see figures 9 and 10).

Fig. 9 and 10. Examples of suburban agglomerates in Coimbra, Portugal

Generally private transport modes, normally motorized but sometimes cyclist, by themselves or integrated in a multimodal solution, will tend to have a dominant role in the service of the accessibility to these spaces. Public transport systems will tend to be reserved to the more specific, but not less important, roles related with guaranteeing minimum accessibility conditions to everyone, thus fulfilling what is normally considered to be public service. When the problem is the connection of these zones with urban city centers, then standard public transport solutions can be competitive although, when lower density areas are concerned, multi-model solutions like P&R tend to be a better choice.

Inside the different agglomerates mobility should be well served by pedestrian networks which must also guarantee good access conditions into public transport stops. The quality of service provided by these pedestrian networks is very much dependent on the way conflicts with the road network are dealt with, with the control of the trough motorized traffic speeds, eventually using traffic calming solutions, being an important aspect.

7. Integrated Strategies and Measures for an Efficient Urban Mobility

7.1 The Need for integrated Strategies

The urban transport systems' possible intervention strategies are quite varied and require a coordinated development and implementation.

Generally, as seen before, the general aims of a transport policy are to better serve a reduced number of private car users, while at the same time more people is convinced to use interesting, more sustainable modes, like public transport, bike or foot.

The different relevant strategies generally involve:

- The optimization of the road network performance;
- The introduction of car usage restrictions, particularly in the most sensitive areas;
- The creation or improvement of competitive, sustainable alternatives (public transport, bike or pedestrian systems) and their promotion;
- The intervention at the land use level in order to change the urban mobility patterns so that the more sustainable modes can be more competitive.

7.2 Optimization and Restriction of Private Car Usage

As was referred before the private car, within the urban mobility context, tends to be the most inefficient mode of transport at both the energy and environment levels.

The intervention strategies in relation with this mode tend to include three different focus areas: first, improvement of the infrastructure operational efficiency so that more vehicles can better use the available infrastructure or, preferably, that the same or even less number of vehicles is better served by a smaller infrastructure; second, improvement of the usage of the available transport capacity by improving the vehicles' occupancy; third, promoting the shift to other, more efficient, modes by introducing a coherent set of restrictions towards the movement and parking of private cars, particularly in the most sensitive urban areas.

In the first group of measures one can identify several ITS based ones, such as the implementation of Centralized Real Time Traffic Control and Traffic Information Systems, through which it is possible not only to optimize available capacity but also to influence the way in which the users use the infrastructure.

The second group includes interventions not only at the infrastructure operation level but also through information and promotion actions.

At the infrastructure level the basic measure relates with the implementation of an integrated network of High Occupancy Vehicles (HOV) lanes which are to be used only by cars with 2, or eventually 3 or more occupants, thus optimizing the use of the existing road capacity.

This kind of measure should be complemented with the implementation of Car Pooling solutions, where two or more persons, who have similar mobility needs, choose to use the same vehicle, normally sharing their costs, and thus reducing the number of vehicles circulating. This can be done by simple promoting actions or by the creation or support of structured car pooling systems and companies which try to speed up and optimize the trip matching processes.

The promotion of a less intense use of the private car can be done by physically restricting the access, the movement or the parking in the most sensitive areas, for example by reducing the capacity or speeds provided by the road network or by reducing the parking offer or by introducing timing restrictions.

The intervention can also be done by increasing the cost of using the private car either by increasing the parking tariffs or by the implementation of urban tolls. This last type of solution has been gaining popularity and proved to significantly contribute to control access levels and conditions to sensitive areas (Commission of the European Communities, 2006).

Other interesting instruments are Car-Sharing schemes, where people give up the ownership of the car and is encouraged to use cars, rented in some kind of pool system, just when that is absolutely necessary. This kind of scheme is particularly attractive in areas where parking is very difficult or expensive.

7.3 Promotion of the Pedestrian and Bike Modes

The pedestrian system is of central importance in the implementation of any integrated sustainable transport policy. In fact the pedestrian mode has all the potential to be the main mode in city centers, in residential ones or, in any sensitive locations in general.

Intervention on the infrastructure must be designed in a coordinated way in all its components (circulation links, intersections with other sub-systems and mode interfaces), with the clear perception that its quality of service and competitiveness depends decisively of the existence of direct, continuous, effortless, comfortable and safe circuits connecting all the important trip generators. For that to be possible a number of infrastructure intervention methodological principles should be considered:

- Separation of the road and pedestrian networks particularly at the major road links, in order to guaranty high levels of pedestrian safety and to contribute to create attractive circuits;
- Creation of a dense pedestrian network connecting not only the trip generators but also connecting all the existing modal interfaces;
- To attend with special care the specific needs of the most vulnerable pedestrians such as children, old people and people with reduced mobility.

In what concerns the promotion of the bicycle transport mode it should be noticed that it depends in first place on the existence of a quality bicycle specialized supporting infrastructure, which includes not only a network of bicycle paths, but also involves parking areas located near the major trip destinations and public transport interfaces, as well as other complementary equipments such as specialized maintenance service companies and, desirably, sanitary installations with showers in the working places and schools.

In what concerns the implementation of a quality bike network the design principles have a lot in common with those presented in relation to the pedestrian networks (dense networks providing direct, continuous, comfortable and safe connections between all major trip generators and mode interfaces), although consideration must be given to the bicycle mode special needs relating mainly with the difficulty it has to deal with particularly steep gradients.

The promotion of the pedestrian and bike modes should go beyond the investment in the infrastructure, with the implementation of awarness campaigns, aiming to change the negative stereotyped image that, in many countries and communities, still is associated with these modes. In particular, special campaigns, focused on the promotion of the bicycle use, with family involvement but particularly directed at small children and teenagers, are promising avenues. Also innovative initiatives developed at important working places and schools have also shown to have some potential (Commission of the European Communities, 2006).

7.5 Promotion of Public Transport Modes

Public Transport (PT) promotion involves the creation of higher density services, particularly those with no vehicle interchanges within the same trips, along extended periods of time of every day, and with a good quality of service which involves reliability, speed, comfort and competitive pricing.

Competitive PT also implies the adoption of a door-to-door service approach as opposed to a stop-to-stop approach.

A basic element of the intervention strategy must be the creation of an efficient circulation supporting infrastructure where the existence of reserved segregated paths is essential to guaranty reliability and good commercial speeds. These segregated paths should be as continuous as possible with that being an obligation for the heaviest rail based systems.

For the non-completely segregated solutions the assumption of priority awarded at the crossings with other modes' networks is also very important and the highest capacity the specific PT mode has, the more this priority should be given in a systematic way. This can be made through the implementation of Automatic Vehicle Location and Identification (AVL) integrated with Real Life Traffic Control (UTC) Systems.

Another potentially interesting measure is the creation of more direct circuits than those allowed to the private car mode in order to create a positive discrimination for the PT. This can, for example, be achieved by the implementation of counter-flow Bus lanes.

The competitiveness of the Public Transport systems also depends very much on the quality of the location and intrinsic quality of their stops, on the existence of very good connections, particularly with the pedestrian paths, and on the existence of good accesses for users with reduced mobility.

One other important system element is the existence of a good information system, capable of providing information over schedules and service connections both remotely, in the stops and inside the vehicles, and whenever possible, providing it in real time.

7.6 Promotion of Multimodal Solutions

As explained before multimodal solutions present the potential to capture a significant number of important types of urban trips.

For that to be possible adequately designed and located modal interfaces are essential components in order to counterbalance the inevitable shock which results from the need to change mode and or service in the middle of the trip. In these points a panoply of different components, not only directly linked to the transport systems but even complementary ones like sanitary installations, media centers or fast food outlets, are of relevance.

Other decisive components are the multi modal integration not only of the existing information systems, dealing with door-to-door information, but also of the ticketing systems.

Finally it is also important to notice that the creation of really competitive multi modal solutions implies the complete integration of the services not only in terms of scheduling but also in what concerns tariffs.

7.7 Land Use Planning Complementary Intervention Areas

The urban structure, namely its type and concentration, influences the mobility patterns, particularly in what concerns the geographical distribution and concentration of the trips.

It has been argued that the "compact city" is the most efficient form of organization from an energy perspective, also presenting some potential social and economic advantages.
Other urban structures, such as that of "decentralized concentration" or that of the "linear city" are also considered efficient from a mobility perspective (Stead, 2001).
As it is understandable the basic characteristic common to all these types of structures is the existence of high urban concentrations around the points with high accessibility (Banister, 2007). This gives a significant potential for public transport competitiveness since it enables significant efficiency gains which, for example at the energy consumption level can represent 10-15% reductions of transport fuel usage by comparison with other less efficient urban structures (Ecotec, 1993).
In parallel, the wider implementation of mixed-use urban developments pursuing a local self-sufficiency logic, which tends to reduce the need for long, complex trips, is considered to be an interesting urban planning option.
From what was presented here it is apparent that, in the medium-long term, the assumption of land-use planning options consistent with sustainable mobility models, can have a significant impact and thus contribute towards reducing the current private car mode over dependency.
In parallel, the recent technological developments relating with information and communication technologies (ICT) are creating opportunities to improve urban mobility conditions by impacting on people mobility needs and behavior.
In fact their use presents the potential for greater human activities' scheduling flexibility, with reduction of peak hours travelling, and even trips' elimination. However, these kinds of impacts are complex and not yet completely quantifiable (Banister and Stead, 2004).

8. Sustainable Mobility Benchmarking Case Studies

In the following points a structured, but resumed presentation is made of a number of international benchmarking real life examples, representing different integrated and efficient intervention strategies.
In the presentations an effort is made to identify the basic intervention principles associated with each of the basic options, while at the same time some attention is given to the actions and measures used to implement the policies.

8.1 Public Transport Systems' Promotion

(a) KARLSRUHE – Germany (implementation of Tram-Train system)

Karlsruhe, is a city with around 273,000 inhabitants which is surrounded by a 1.3 million inhabitants region.
Since 1961 the urban and sub-urban rail based systems have been jointly organized and managed, covering a network with over 600Km. The objective of this merger was to better integrate both systems (Lehmbrok et al, 2007), with the final objective of creating a more competitive public transport system capable of attracting previously private car trips.
Over time the adopted management model assumed a number of basic strategic options:

- Integration of the tram and train lines avoiding unnecessary transfers between the services provided by the two networks;

- Construction of new stretches of lines connecting both networks, adaptation of the interfaces and acquisition of new rolling stock capable of using the two different electrical propulsion systems;
- Introduction of a number of new urban and sub-urban stations taking advantage of the higher acceleration and deceleration capability of the rolling stock;
- Integration of the two services scheduling and increase of the services' frequency as well as better coordination with the road based public transport services and with the private car and the bicycle through new Park&Metro and Bike&Metro systems;
- Integration and development of the information, ticketing and tariffs systems;

In parallel the densification of the urban developments served by the tram-train system was pursued in this way increasing the potential number of users.
The population has been always informed and involved in the project through newspapers, magazines and pamphlets based campaigns.
The results from this project have been extremely interesting with the system demand growing 400% from 1992 to 2000 to around 150 million passengers per year of which 40% were previous private car trips.

(b) STRASBOURG – France (integrated transport system)

Strasbourg is a medium size French city with around 250,000 inhabitants. Since the nineties the sub-urban areas quick development associated with an accelerated concentration of the commerce and services in the city center have aggravated significantly the mobility problems.
It was then decided to assume a more sustainable transport policy through the promotion of the more environmentally friendly modes in detriment to the private car mode, coordinated with a land use policy directed at the development of a compact city guarantying short distances between the major traffic generating and attraction areas.
The transport system restructuring was based on the implementation of a new urban tram system associated with its efficient coordination and integration with a restructured road based public transport system (Lehmbrock et al., 2007; Difu, 2007).
In parallel, new multimodal Park&Ride and Bike&Ride solutions were developed near the more suburban tram stations, particularly the terminal ones, and bicycle use promoting campaigns were developed.
Special care was given to the urban integration and image of the new tram system having in consideration the special needs of the nobler city center spaces (see Figures 11 and 12), as well as to the accessibility needs of people with reduced mobility. Taking advantage of the opportunity given by the implementation of the new tram system, significant city center spaces were made pedestrian areas.
At present the tram network is constituted by four lines totaling 31.5Km and directly serving 70,000 inhabitants. This network offers high service frequencies throughout the day. The tram network is complemented by a bus network with a total length of 310Km in the urban areas and 280Km in the suburban ones.

Fig. 11 and 12. Areas of the City Center of Strasbourg in France

It should be noticed that the all project was initially received with some scepticism by residents and commerce. In order to overcome it the local authorities have developed a comprehensive information and promotion campaign based namely in public information events.

Following the implementation period and with the beginning of the new tram operation, the scepticism has disappeared and the system became a significant success resulting in a 32% public transport passenger increase from 1992 to 1995, with a parallel reduction of private car use of 17%, with the total number of annually transported passengers in 1996 being 41.9 million (Lehmbrock et al., 2007).

Amongst the residents the creation of the new pedestrian areas in the city center, made possible by the implementation of the new transport system, was viewed as a major contribution to the local urban quality of life.

8.2 Pedestrian and bicycle promotion

Many cities around the world have over the last few decades assumed a coherent strategy of systematic promotion of the bike and pedestrian modes as real alternatives to the use of the private car. Good examples are amongst other Odense and Copenhagen in Denmark, Munster and Berlin in Germany, Bolzano in Italy, Amsterdam in The Netherlands, Sandnes in Norway, Barcelona in Spain, Basel in Switzerland or Davis in the USA.

(a) ODENSE - Denmark (bike promotion)

The Danish city of Odense, which has 145,000 inhabitants, in 1999 has received the "National Cycle City of Denmark" awarded by the Danish Transport Ministry (Adonis, 1998) as a recognition to the systematic and coherent implementation of actions and measures to increase bicycle use.

The city has developed a large scale implementation which have involved 60 case studies financed at the national and local level and involving investments of over 3.5 million Euros.

The city has assumed the objective of obtaining a dominant use of the bicycle in the access and mobility in both the city center and its surrounding areas, in order to preserve the traditional urban space quality and attractiveness, while at the same time the city accessibility should be improved.

A 512Km long bicycle network has been built representing the basic infrastructure component of the system, while at the same time complementary actions such as safety promotion (intersection priority measures, speed controlled corridors, traffic calming measures), adaptation of the transport regulatory system, creation of new specialized services and information systems, were also carried out.
Special attention was also given to the development of promotion and information actions directed not only to the population in general but, at the same time, electing school children as a basic target in relation to home-to-school trips.
All these actions were supported by a strong marketing campaign, in which a special logo and an informative magazine were created.
All the implementation was also subjected to systematic monitoring procedures, in order to guaranty the quality of the solution with immediate correction of any identified deficiencies and regular maintenance of the infrastructure.
The population has assumed the project in an extremely positive way, so that by 2002 the bicycle was already the most important mode used (43.1%), followed by the pedestrian (23.2%), the private car (21%) and public transport (14%) modes.

(b) COPENHAGEN - Denmark (bike and pedestrian modes promotion)

Copenhagen is a 1.15 million inhabitants city with a very densely populated (5,700 hab/Km2) city center.
Before the implementation of the new integrated transport policy the city, which possesses a limited road network capacity, presented significant levels of congestion which were severely affecting the public transport level of service.
To address these problems local authorities decided to promote the bike and pedestrian modes and, at the same time, to introduce private car restrictive measures. This was carried out accompanied by permanent involvement and sensitization of the local population.
Many roads were intervened with the objective of creating shared quality public spaces, where priority was given to pedestrians and bicycles over cars. Others were completely devolved to pedestrians and bicycles.
All these places were object of urban renewable processes where, in parallel with the implementation of traffic calming measures, other interventions were carried out based on the substitution of bituminous pavements by other aesthetically more appealing, and placement of new urban furniture.
In parallel, a comprehensive set of bike supporting actions and measures were implemented ranging from explanatory campaigns, focused on particular segments of the population (immigrants, women, elderly, children, ...), to physical measures (lengthening of the bicycle network, implementation of intersection bike priority, introduction of counter-flow bicycle lanes, ...), to the implementation of a free availability of bicycles in city scheme.
New bicycle acquisition financing programs were also implemented, particularly directed at private companies, who then would devise their own logos and imaging on the bikes not only with marketing but also theft prevention purposes.
All the intervention was implemented progressively with the permanent participation of the residents and other stake holders, namely during the development of the relevant projects, in order to guaranty their acceptance off all the program.

Significant improvements at the quality of life level were obtained, reflected namely in the improvement of the noise and pollution levels and making the intervened spaces attractive and safe (Adonis, 1998).
The population has adopted the policy and that is reflected in a 2004 modal split that was already clearly dominated by the soft modes and by public transport with the private car reduced to less than 30% of the modal share.

8.3 Intervention at the Land Use Level

BOCHOLT – Germany (compact city)

Bochold is a very compact city where the basic mobility controlling intervention strategy has been the maintenance of small trip lengths, compatible with the bicycle and, specially, the pedestrian mode. The basic objective of the different actions is to maintain a situation where 90% of the population lives within less than 3Km of the city center.
This is achieved based on strict local regulations concerning land use rules, specially directed to the location of new commercial developments, where licensing is awarded only when the essential of the respective accessibility can be guaranteed on foot or bicycle.
Also a strong emphasis is put on the achievement of significant diversity of land use types in the different city neighborhoods in order to reinforce each area functional autonomy and thus contributing to contain and even reduce average trip lengths.
In parallel, car usage restrictions were imposed, ranging from the limitation of circulating speeds in residential areas by implementation of 30 Km/h zones, to the reinforcing of parking restrictions with the imposition of parking fees covering the entire city center.
As a result the city adopted the use of the bicycle, being at present one of the German cities where bicycle mobility is more important with a 35% modal share, and justifying the qualification of a "cycling city". This has been accompanied with the maintenance of a very high local quality of life for its citizens.

8.4 Sustainable Mobility in Developing Countries

Poorer countries, although possessing much lower motorization levels, are a significant cause for concern because of the potentially catastrophic environmental and energy impacts of their mobility conditions not only at the present, but particularly in the near future, due to their aging and outdated present motorized vehicles' fleets and, specially, to the potential for a very quick grow of their numbers.
In this context the World Bank has been developing new grant programs to help these countries to fight the climatic changes causes through the adoption of solutions capable of reducing pollution emissions while at the same time creating more efficient mobility conditions.
Within this line of action it is worth referring to a set of measures which, for some time already, have been under development in the Brazilian city of Curitiba.
This intervention program involves the transformation of the major federal highway (BR-116/476), which crosses through Curitiba, in an urban avenue served by a new high capacity Train Line ("Trem Urbano"), which will be closely interconnected with the city extensive public transport network. Furthermore this project contemplates other interventions such as the construction of new leisure parks and cycle paths as well as

remedial interventions directed at road traffic safety. It also involves intervention at the land use intervention level.

This program should enable the reduction of average trip duration, in parallel with improvements in the population accessibility to an improved transport system and, potentially, enabling a reduction of the metropolitan area mobility operational costs.

In general one of the main rules which might be defined for the implementation of new more sustainable mobility policies in developing countries cities is the need for the selection of low cost solutions which can be progressively implemented and evolve through time.

In first place, due to its importance, it is worth referring to the need for the progressive and systematic implementation of quality and dense pedestrian networks. In this respect particularly relevance assumes not only the construction of pedestrian paths but, specially, the improvement of their intersections with the road network.

A strong effort in creating good conditions to the use of the bicycle also seems to present significant potential due to the reduced costs involved, not only in the construction and maintenance of the infrastructure but, particularly, on the costs of the vehicles. A coherent and systematic investment in the creation of bicycle paths along the major traffic corridors tends to be very important.

Finally, particularly in the bigger cities and metropolitan areas, it is essential to progressively create dense networks of public transport services' supporting infra-structures. A specially promising strategy might be the identification and progressive implementation of exclusive "Busways", which at first can be used by the more traditional forms of public transport (as are for example the "Chapas" from Maputo in Mozambique), and which later on can be used to implement real Rapid Transit networks, which can be permanently road or rail based or can evolve from one type to another over time as demand and wealth grows.

9. References

Adonis (1998). *Best practice to promote cycling and walking,* Analysis and Development Of New Insight into Substitution of short car trips by cycling and walking

Austroads (2003). *Valuing Environmental and Other Externalities,* Report - AP-R229, Ed: Austroads.

Banister, D. & Stead, D. (2004). Impact of information and communications technology on transport, *Transport Reviews* 24 (5), 611–632

Banister D. (2007). The sustainable mobility paradigm, *Transport Policy* 15 (2008) 73–80

Civitas (2006). Sustainable Urban transport, *Final report from de European project Trebsetter,* Anna Hadenius, Inregia; Jonas Ericson, Environment and Health Administration, City of Stockholm

Commission of the European Communities (2001). *Livro Branco - A política Europeia de transportes no horizonte 2010: a hora das opções,* Comissão das Comunidades Europeias, Bruxelas

Commission of the European Communities (2006). Keeping Europe moving: sustainable mobility for our continent, *Midterm Review of the European Commission's 2001,* Transport White Paper, COM (2006) 314 Final, 22 June 2006, Brussels

Difu (2007). *Sustainable Urban Transport and deprived urban areas Good Practice Examples in Europe,* German Institute of Urban Affairs (Difu)

Ecotec (1993). *Reducing Transport Emissions Through Planning*, HMSO, London

EEA (2000). *EMEP/CORINAIR Atmospheric Emission Inventory Guidebook - Revised version*, Denmark, Expert panel of the UNECE/EMEP Taskforce for emission Inventories

Infras/IWW (2000). *External Costs of Transport: Accident, Environmental and Congestion costs in Western Europe*, Ed: International Union of Railways - UIC

Infras/IWW (2004). *External Costs of Transport - Update Studt*, Ed: International Union of Railways - UIC

Lehmbrock M.; Spott M. & Beckmann K. J. (2007). *Sustainable Urban Transport and Deprived Urban Areas, Good Practice Examples in Europe*, German Institut of Urban Affairs, Berlin

Stead D. (2001). Relationships between land use, socioeconomic factors, and travel patterns in Britain, *Environment and Planning B: Planning and Design* 28(4) 499 - 528

Vuchic V. R. (2007). *Urban Transit Systems and Technology*, Ed : John Wiley & Sons, Inc

4

Urban Flood Control, Simulation and Management - an Integrated Approach

Marcelo Gomes Miguez, Luiz Paulo Canedo de Magalhães
Federal University of Rio de Janeiro (UFRJ)
marcelomiguez@poli.ufrj.br, luizpaulocanedo@hotmail.com
Brazil

1. Introduction

History shows that places near rivers were attractive as sites for ancient civilisation development. Mesopotamia, China, Egypt and Rome are some examples. Fertile lands, transportation and water supply were main factors. During the industrial revolution, however, stormwater started to be considered as a great matter for urban life. Urban floods started to increase in magnitude and frequency. Street gutters were used to convey stormwater and wastewater as well. Diseases spread around the industrial city and waters had to be discharged as faster as possible. As a consequence, urban drainage started to play an important role in cities life.

It is possible to say that urbanisation is an inexorable trend. The urban population has been increasing significantly in the last two centuries, since industrial revolution took place. The consequences of this process incurs in great changes of the natural environment. Urbanisation process tends to substitute natural vegetation for impervious surfaces, thus reducing infiltration. It also tends to eliminate natural detention ponds, to rectify river courses, among other actions, that greatly interfere with superficial flows. In general, floods in urban areas present greater runoff volumes and flow velocities, resulting in higher flow peaks and water stages. This way, urbanisation aggravates floods and, as it could be seen in cities development, it was not always possible to accomplish urban growing with the adequate infrastructure, especially in developing and poor countries. Even in wealthy countries, urban growth stresses the existing infrastructure.

Urban floods disrupts social systems and cause significant economic losses. Among the impacts produced, there are health hazards and losses of human lives, flooding of housing, commercial and industrial properties, flooding of streets and intersections, causing traffic delays, disruption of services such as water supply, power supply and sewerage.

Flood control is, thus, one of the major issues with which urban planners must deal nowadays, once floods play a dramatic role in the cities. Additionally, the lack of planning frequently worsens this situation. Many times, the absence of systemic design tools capable to represent the problem in an integrated approach leads to a decision process in which local solutions may be inadequate for the whole system needs. An important tool to be considered refers to the mathematical modelling of hydrologic and hydraulic processes.

The concepts applied to stormwater control measures design have changed a lot in the past decades. The traditional approach focused on the drainage net correction, by canalising and rectifying watercourses, in order to improve conveyance. More recent developments tend to search for systemic solutions. New concepts focus on flood risk management aspects, concerning a multidisciplinary approach that considers aspects of prevention, mitigation and recovery of the hazard prone area. Cities are faced with the challenge to find a sustainable way in order to equilibrate harmonic growing with built environment.
In this context, the aim of this chapter is to present a comprehensive and up-to-date review on issues related to flood control and mathematical modelling, integrated with urban planning policies and strategies.
The topics covered by this chapter comprise a general frame of urban drainage problems and their interaction with urban planning; a basic review on historical aspects of the evolution of urban flood control; a presentation of structural and non-structural flood control measures, including modern sustainable drainage techniques; and a broad discussion on hydrologic and hydrodynamic urban flood modelling techniques, illustrated with some case studies applied to the State of Rio de Janeiro, Brazil.

2. Urbanisation and Floods

Floods are natural and seasonal phenomena that play an important environmental role. However, human settlements interfere with flood patterns, majoring their magnitude and frequency of occurrence, turning higher the associated level of risk regarding people, buildings and economic activities. Urban floods range from localised micro-drainage problems, inundating streets and troubling pedestrians and urban traffic, to major inundation of large portions of the city, when both micro and macro-drainage fail to accomplish their basic functions. These problems can lead to material losses to buildings and their contents, damage to urban infrastructure, people relocation, increased risk of diseases, deterioration of water quality, among others.
Considering it in a simple way, when rainfall occurs a portion of the total precipitation is intercepted by vegetal canopy or retained at surface depressions, another part infiltrates and the rest of it flows superficially over the terrain, conveying to channels and lower areas. The main modification introduced by the urbanisation process to the water budget refers to an increase of superficial runoff production, as can be seen in figure 1. Table 1 summarises the different impacts of urbanisation over a river watershed. Studies held by Leopold (1968) showed flood peaks majored about six times, when compared to floods in natural conditions.
The fact that must be faced is that the city can influence runoff pattern changes and the state of ecological systems not only within itself but also in the whole river system downstream, including its surroundings. This fact, historically, resulted in shifting the traditional conveyance approach in stormwater management, during the 1970s, to the storage approach with a focus on detention, retention and recharge. Later on, the evolution of this concept, during 1980s and 1990s, made stormwater to be considered as a significant source of pollution, and the goals of stormwater management shifted again in order to protect natural water cycle and ecological systems by the introduction of local source control, flow attenuation measures and water quality treatment systems such as retention ponds, wetlands and others (Niemczynowicz, 1999).

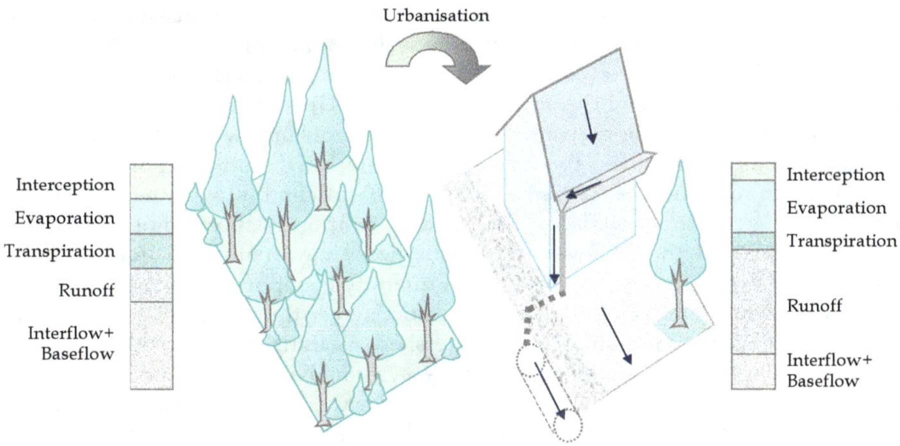

Fig. 1. Schematic picture of urbanisation changes in the water balance

Causes	Effects
Natural vegetation removal	Higher runoff volumes and peak flows; greater flow velocities; increased soil erosion and consequent sedimentation in channels and galleries.
Increasing of imperviousness rates	Higher runoff volumes and peak flows; less surface depressions detention and greater velocities of flow.
Construction of an artificial drainage net	Significant increasing of flow velocities reduction of time to peak.
River banks and flood plain occupation	Population directly exposed to periodic inundation at natural flooded areas; amplification of the extension of the inundated areas, as there is less space to over bank flows and storage.
Solid waste and wastewater disposal on drainage net	Water quality degradation; diseases; drainage net obstruction; channel sedimentation

Table 1. Urbanisation impacts over floods

Flood control concepts are evolving continuously, accompanying historical demands of urbanisation and its consequences. When a city starts to grow near a river, at a first moment, this city can only be inundated in extreme events, when natural floods occupy larger portions of floodplain. Urbanisation, however, changes landscape patterns, aggravating floods by increasing surface runoff flows. In this way, floods become greater in magnitude and time of permanence, occurring even more frequently.

The traditional approach for this problem focused on the drainage net itself, arranging channels and pipes in an artificial flow net system, with the objective to convey the exceeding waters away from the interest sites. At this initial moment, the canalisation solution is able to deal with floods in a certain area, transferring waters downstream with no major consequences. As time passes, urbanisation grows and more areas of the watershed turn impervious. Upstream development stresses the system as a whole and the drainage net fails once again. By this time, it becomes difficult to depend exclusively on improving channels conveyance capacity to try to adjust the system behaviour.

Urbanisation itself limits river canalisation enlargement. Streets, buildings and urban facilities now occupy banks and the original flood plain. Upstream reaches of the main river cannot be canalised without aggravating downstream problems, where the former city area lays. Focus now must be moved to a systemic approach, where the whole basin must be considered. Distributed actions spread around the basin comply with the drainage net in order to control generation of flows. Spatial and temporal aspects must be considered together in a way that the proposed set of solutions may reorganise flow patterns and minimise floods. In this context, not only water quantity is important, but also water quality is an issue to be considered. Distributed interventions over the urbanised basin can also act on the control of diffuse pollution from watershed washing. Here arises the concept of sustainable drainage, which states that drainage systems have to be conceived in order to minimise impacts of urbanisation over natural flow patterns, joining quantity and quality aspects, meeting technical, social, economic and political goals, without transferring costs in space or time.

In order to illustrate the interaction between urban development and flood control, as discussed above, table 2 pictures a schematic frame of a hypothetical basin urbanisation process. Knowing the sequence of facts presented in this table, it is possible to say that it would be easier to imagine another course of actions, working in a preventive way and avoiding undesirable flooding. Planning in advance, mapping of flood hazard prone areas, developing environmental education campaigns, establishing adequate legislation, in order to restrict runoff generation, among other measures, would configure a set of procedures that could allow a rational coexistence of human settlements with natural floods.

However, it is impossible to prevent everything, as it is impossible to go back in time. The historical aspects of urban development lead to all sort of established situations, where urban floods occur. There is not one best answer for this problem. Each basin has to be considered with its own characteristics, particularities and historical background, once the diversity involved may arise lots of differences from case to case. However, many studies have been developed in order to propose new concepts and alternatives.

Macaitis (1994) edited a book for American Society of Civil Engineers, where it is presented the concept of urban drainage rehabilitation. This book showed a series of studies that focused on identifying urban drainage functioning, defining maintenance procedures and proposing complementing structures (as ponds, by-passes, flood-gates, etc), in order to allow system operation to minimise flood impacts. Hunter (1994), in a paper presented at this book stressed that it is important to maintain channel conveyance capacity, by treating flood causes and not its consequences. A drainage system working as designed can be able to sustain nearby communities safety and health.

Coffman et al. (1999) proposed a design concept of low impact development (LID). LID design adopts a set of procedures that try to understand and reproduce hydrologic behaviour prior to urbanisation. In this context, multifunctional landscapes appear as useful elements in urban mesh, in order to allow rescuing infiltration and detention characteristics of the natural watershed.

In a similar way, recent trends involve the use best management practices (BMP) in drainage systems design. Best management practices work in a distributed way over the watershed, integrating water quantity and water quality control.

<table>
<tr>
<td>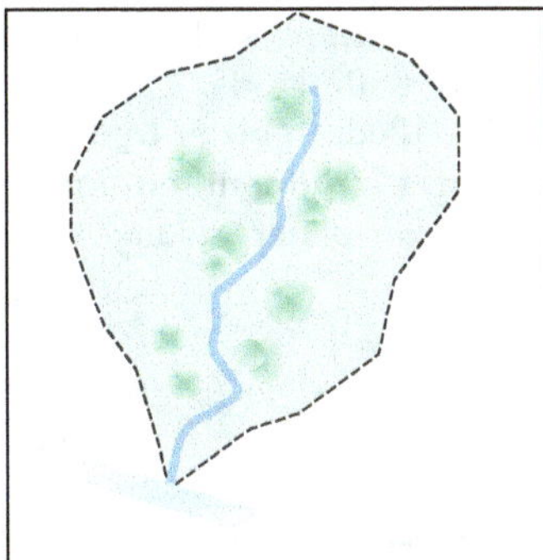</td>
<td>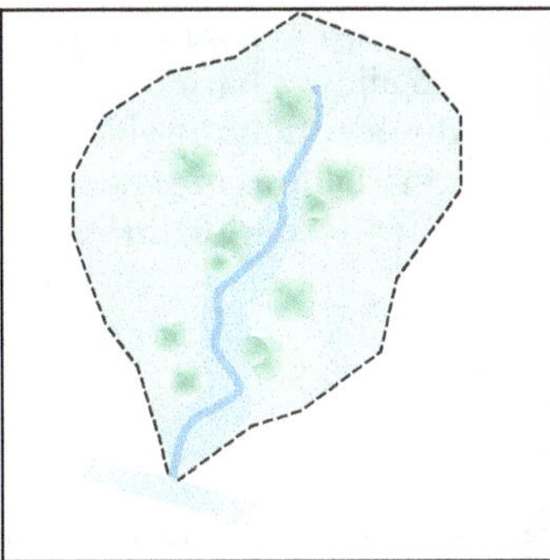</td>
<td>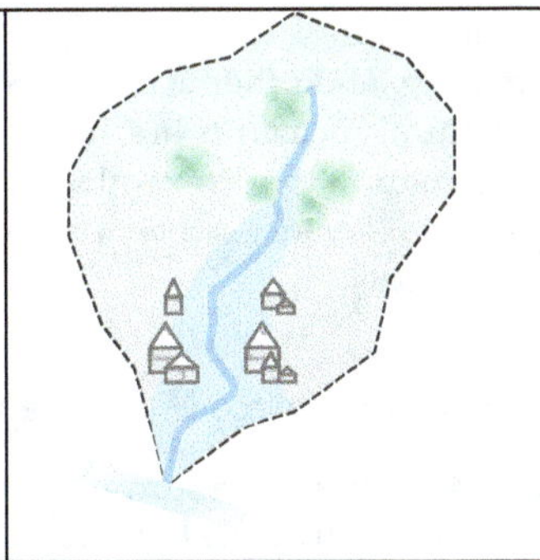</td>
</tr>
<tr>
<td>Natural watershed, with its original land cover, without any occupation.</td>
<td>Natural floodplain.
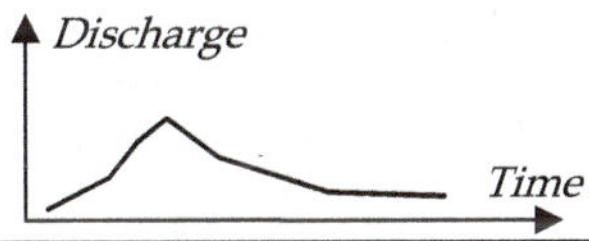</td>
<td>Initial urban settlement: runoff and peak discharge increase
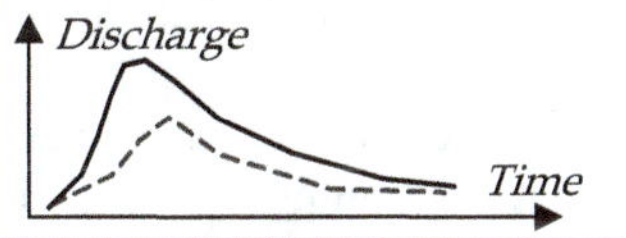</td>
</tr>
<tr>
<td>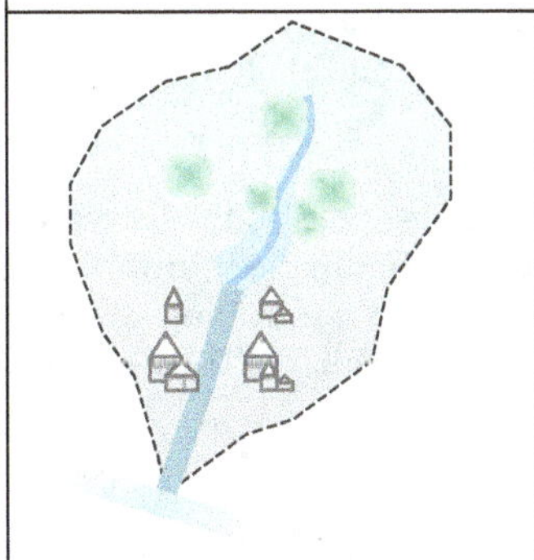</td>
<td>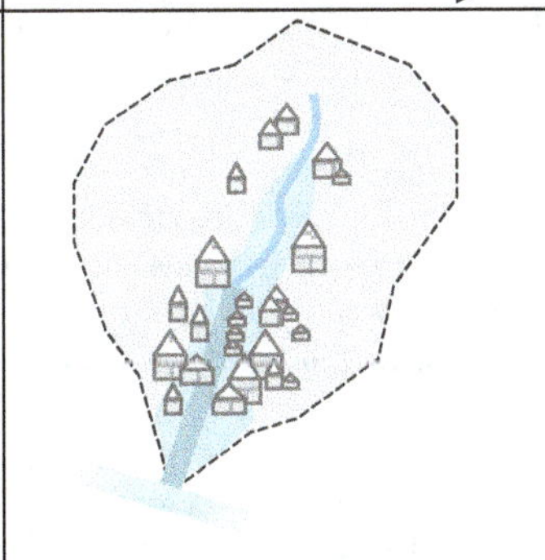</td>
<td>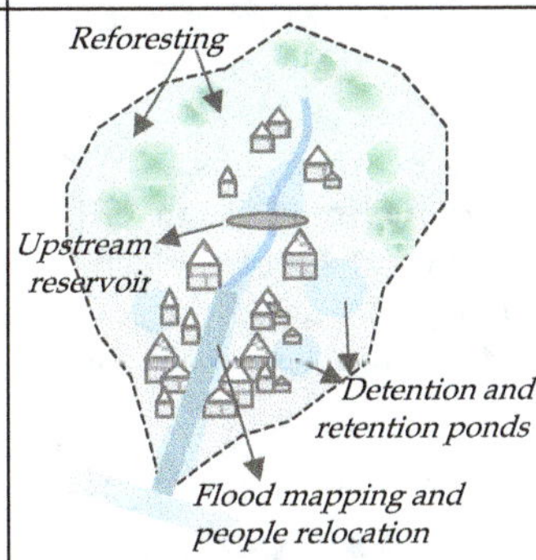</td>
</tr>
<tr>
<td>Traditional approach: canalisation and downstream flood transfer.</td>
<td>Urbanisation growth: greater and generalised floods. Simple focus on channel conveyance does not solve the problem.</td>
<td>Sustainable Drainage: distributed actions over the basin, integrating drainage net and typical urban features, ranging from on-site source control to large structural measures.</td>
</tr>
</table>

Table 2. Schematically evolution of urbanisation and urban drainage solutions

This discussion leads to an important point: understanding how urbanisation interferes with flow patterns is necessary to develop strategies for stormwater management and urban floods control. Urban drainage planning must consider a broad set of aspects and has to be integrated with land use policy, city planning, building code and legislation. It is possible to say that urban flood control demands the adoption of a varied set of different measures of different concepts. Among these measures it is possible to distinguish two greater groups of possible interventions: the structural measures and the non-structural measures. Structural measures introduce physical modifications on the drainage net and over urban basin landscapes. Non-structural measures works with environmental education, flood mapping,

urbanisation and drainage planning for lower development impacts, warning systems, flood proofing, and other actions intended to allow a harmonic coexistence with floods.
Structural measures are fundamental when flood problems are installed, in order to revert the situation to a controlled one. Non-structural measures are always important, but are of greater relevance when planning future scenarios, in order to obtain better results, with minor costs.

3. Flood Control Measures

3.1 Structural Measures

Basically, structural flood control measures compose the most traditional set of interventions on a basin and can be classified as intensive and extensive (Simons et al., 1977). Intensive control measures refer to main drainage net modifications, including river canalisation and rectification, dredging and dike construction, as well as river in line damping reservoir applications, among others. Extensive measures, by their turn, appear spread around watershed surface, acting on source, in order to control runoff generation.
Classical drainage design concepts are intensive methods that focus on improving conveyance. More recent techniques focus on storage and infiltration measures. In the next few lines, some concepts will be presented in order to illustrate flood control alternatives.

(a) Detention Basins

Flood damping is an effective measure to redistribute discharges over time. Increased volumes of runoff, which are resultant from urbanisation, are not diminished, in fact, but flood peaks are reduced. Damping process works storing water and controlling outflow with a limited discharge structure. Figure 2 shows a flood control reservoir (SEMADS, 2001).

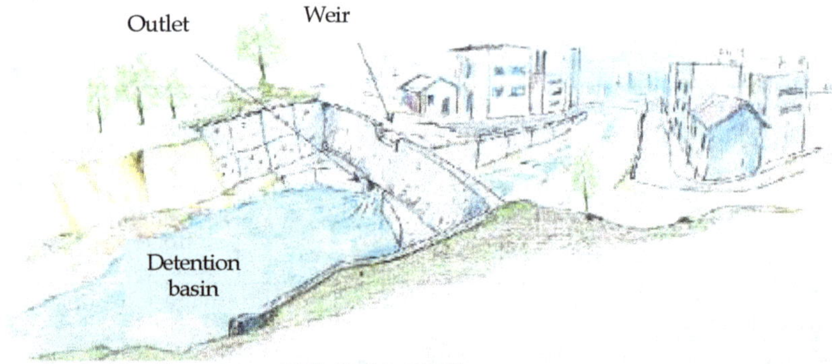

Fig. 2. Detention basin illustration (SEMADS, 2001)

There are several possibilities of application of this kind of measure. Detention ponds may be placed in line with rivers, controlling great portions of the basin, upstream the urbanised area, where occupation is lower and there is more free space to set larger reservoirs. Public parks and squares, as well as riverine areas may be used as detention ponds, opening the possibility to construct multifunctional landscapes (Miguez et al., 2007). Parking lots can also be used, in order to provide temporary storage for flood control. Another possibility, taking into account a smaller scale, on-site detention tanks may be planned as source control measures. Alternatively, it is possible to consider roof detention for the same purpose. In order to illustrate this set of measures, figures 3, 4, 5, 6 and 7 are presented. Figure 3 pictures a reservoir proposed for upper reach of Guerenguê River, in Rio de Janeiro/Brazil, as part of an integrated project of flood control and environmental recovering of the watershed, showing its damping effect (COPPETEC, 2007). Figure 4 shows a detention pond proposed for a public square in Rio de Janeiro/Brazil (COPPETEC, 2004). Figure 5 shows a public square functioning as a multifunctional landscape, also in Rio de Janeiro. It is important to say that this square, called Afonso Pena, was not planned to act this way, but, in practice, when local drainage fails, it acts as a reservoir, avoiding street flooding at its surroundings. Figure 6 shows an on-site detention pond. Figure 7 shows a roof top garden and a roof detention (Arizona, 2003; Woodworth Jr., 2002).

It is important to say that, although providing a local attenuation effect, detention reservoirs must be spatially planned and distributed in an integrated arrangement in order to adequately combine effects for a general positive result.

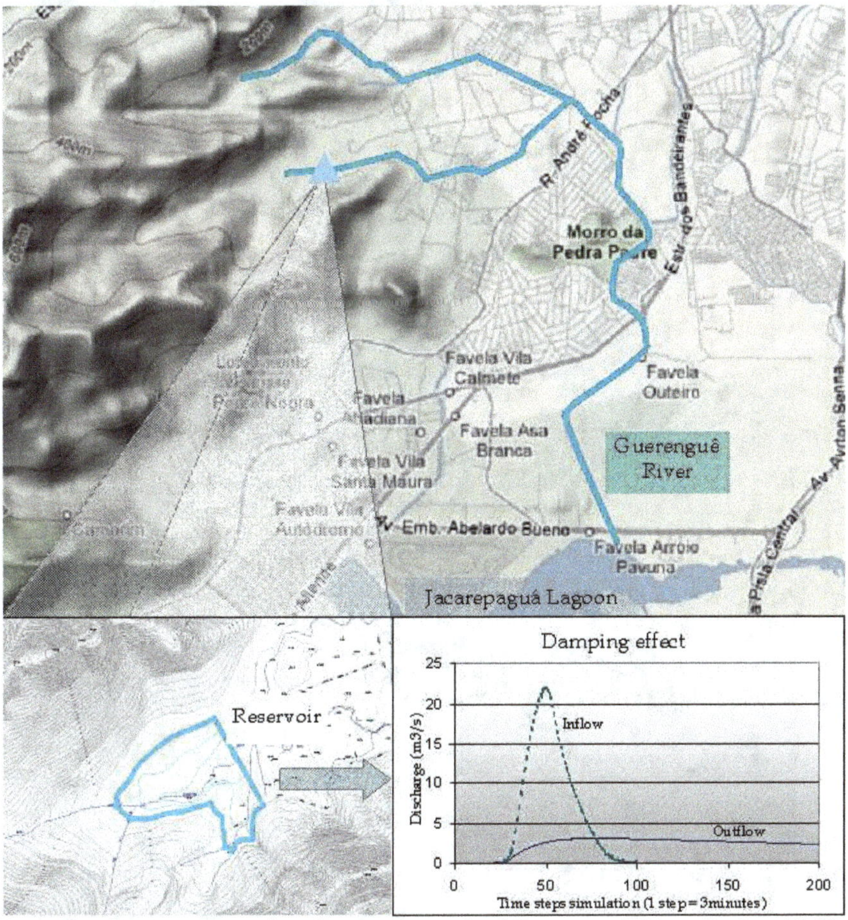

Fig. 3. Detention basin proposed to the upper reach of Guerenguê River Basin – RJ/Brazil

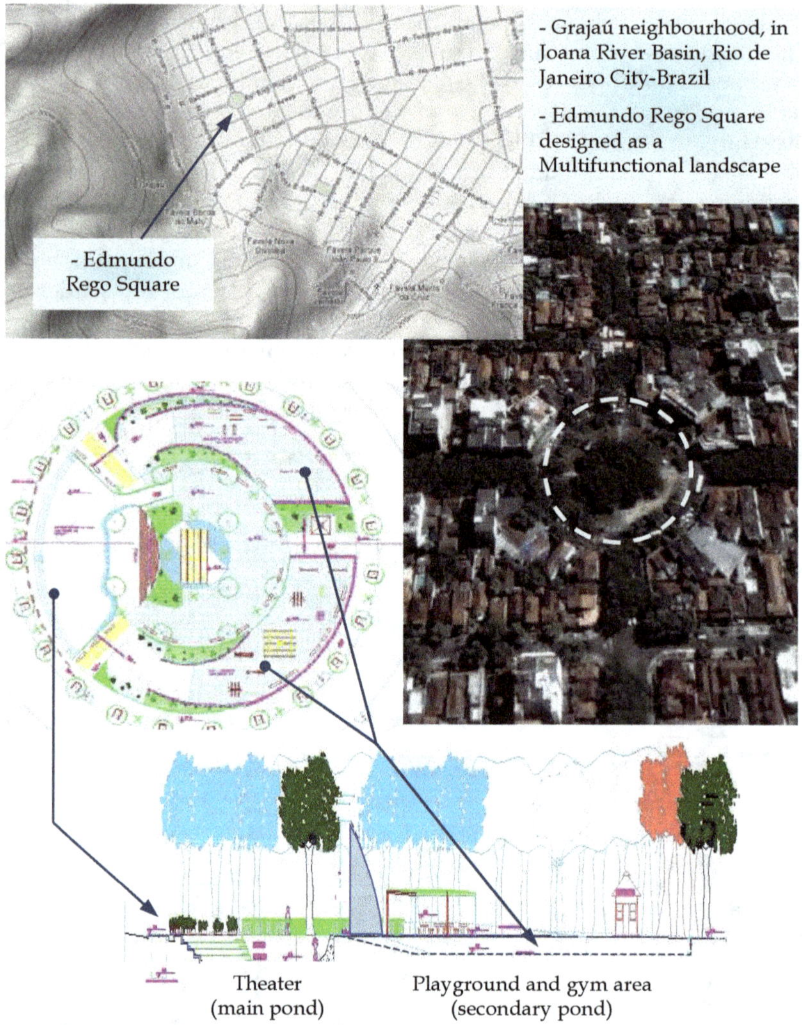

Fig. 4. Edmundo Rego square, at Joana River Basin, designed as a multifunctional landscape

Fig. 5. Afonso Pena Square, acting non-intentionally as a detention pond - RJ/Brazil

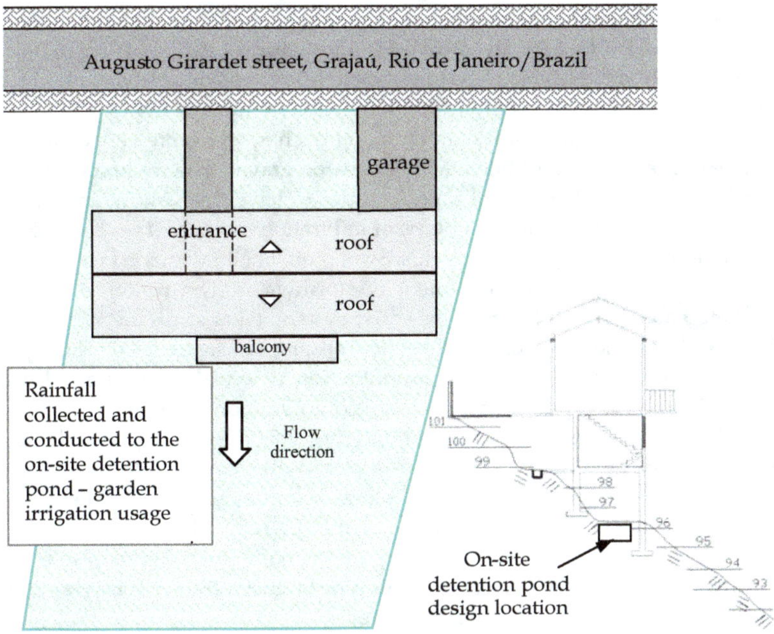

Fig. 6. On-site detention pond, collecting rainfall from the house roof

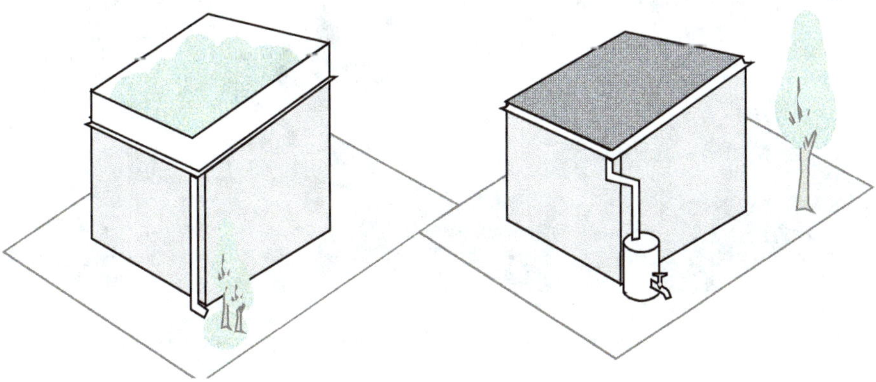

(i) roof top garden, disconnected from drainage net.

(ii) rain barrel, collecting roof top runoff.

Fig. 7. Alternative measures for roof top runoff

(b) Retention ponds

A permanent pool characterises retention ponds. This kind of pond has two main objectives: the first, and most important, is water quality control; the second is water quantity control, although in a minor scale, when compared to the detention ponds. The permanent pool acts allowing the deposition of sediments, helping in diminishing pollutant concentration. Time of permanence of water inside the retention pond is determinant to their efficiency.

(c) Infiltration Measures

Infiltration measures allow to partially recovering the natural catchment hydrologic behaviour. However, it is generally not possible to restore pre-urbanisation conditions, when higher taxes of urbanisation and imperviousness occur. Infiltration measures may be divided into some different categories, depending on how they work. Infiltration trenches, which are very common infiltration devices, are linear excavations backfilled with stones or gravel. The infiltration trench store the diverted runoff for a sufficient period of time, in order to have this volume infiltrated in the soil (AMEC, 2001). Vegetated surfaces are other type of infiltration measure. Two common types of this kind of structure refer to swales and filter strips. Swales are shallow grassed channels used for the conveyance, storage, infiltration and treatment of stormwater. The runoff is either stored and infiltrated or filtered and conveyed back to the sewer system. Filter strips are very similar, but with very low slopes and designed to promote sheet flow (Butler & Davies, 2000). Rain gardens are an especial type of garden designed to increase infiltration potential, presenting also a landscape function. Porous or permeable pavements are a type of infiltration measure where superficial flow is derived though a pervious surface inside a ground reservoir, filled with gravel (Urbonas e Stahre, 1993). Porous pavement upper layer consists of a paved area constructed from open structured material such as concrete units filled with gravel, stone or porous asphalt. Another possibility refers on concrete units separated by grass. The depth of the reservoir placed beneath the upper layer determines the capacity of the measure in minimising runoff. Soil infiltration rates and clogging over time will interfere with the effectiveness of this type of device (Butler & Davies, 2000). Figures 8 and 9 illustrate different types of infiltration measures.

(i)

(ii)

Fig. 8 and 9. Example of rain garden (i) and examples of pervious pavements (ii)

(d) Reforesting

The process of replacing plants in a area that has had them cut down, because of unplanned urban growth, irregular land use occupation or other motives, like economic use of trees, is a very important measure to recover natural flow patterns. Reforestation prevents soil erosion, retains topsoil and favours infiltration. Runoff volumes are reduced and drainage structures keep working efficiently, once a minor quantity of sediments arrives at the system. Renewing a forest cover may be achieved by the artificial planting of seeds or young trees. Figure 10 shows a degraded area in a hill, at Rio de Janeiro City, Brazil, where there was originally a forest reserve.

Fig.10. Degraded hill area – slum occupation substituting a forest

(e) Polders and dikes

The conception of a polder, as illustrated in figure 11, allows protecting a riverine area from the main river flooding, by constructing a dike alongside the channel. Inside the protected area, there are needed a temporary storage basin and an auxiliary channel to convey local waters to this reservoir. Usually, flap gates are responsible for discharging this reservoir when main river water level falls below temporary inside storage water level. Another possibility lays on the use of pumping stations to complement flap gates discharge capacity.

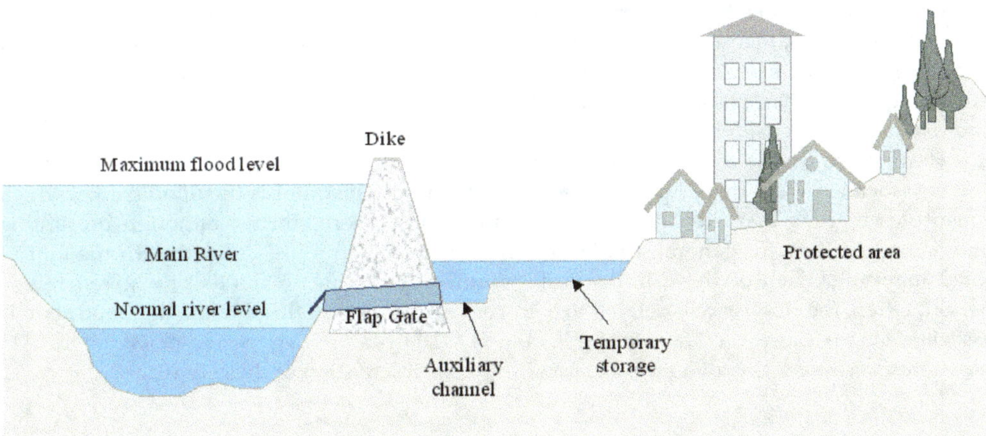

Fig. 11. Illustrative view of a generic polder area

(f) Canalisation

Canalisation is the most traditional measure in drainage works. It is obtained by removing obstructions from riverbed, straightening river course and fixing river banks, resulting in an increased conveyance. Figure 12 shows an example of a canalised river.

Fig.12. Canalised Joana River stretch, in Rio de Janeiro City, Brazil

3.2 Non-structural Measures

Unlike structural works that physically act on the flood phenomena, the aim of non-structural measures is to reduce the exposure of lives and properties to flooding. A wide set of possible actions, ranging from urban planning and zoning to flood proofing of constructions compose this type of measures. The following paragraphs highlight some issues regarding this concept.

3.2.1 Floodplain Management and Regulation

The most important of all non-structural measures is to avoid or restrict the occupation of floodplains. The periodical flooding of riverside areas is a natural process of great environmental relevance. In urban areas, the encroachment of flood plains constitutes a serious problem. The population usually exerts pressure for the occupation of these lands, especially in cases in which there is no recent flooding record or where land use control is ineffective, a common situation observed in poor and developing countries.

Conceptually, floodplain regulation should be based on flood mapping, identification of flood hazard prone areas and establishment of land use criteria. It should also be developed integrated with urban planning activities. In fact, it is extremely desirable that urban zoning and master plans consider aspects related to the regulation of riverine land.

It is common to divide the floodplain into two different zones. The first is called floodway and is associated with areas subject to frequent flooding. The other is the flood fringe, which constitutes regions that may be flooded during more severe storms, although presenting only storage effects. In general, the boundaries of these zones are defined with the aim of flood mapping. Each of these limits is determined according to floods of a given return period. Often, the floodway is related to a 20-year return period flood while the floodplain is associated with more rare events, for instance a 100-year return period flood. Figure 13 illustrates a cross-section of a river basin with the representation of these two zones.

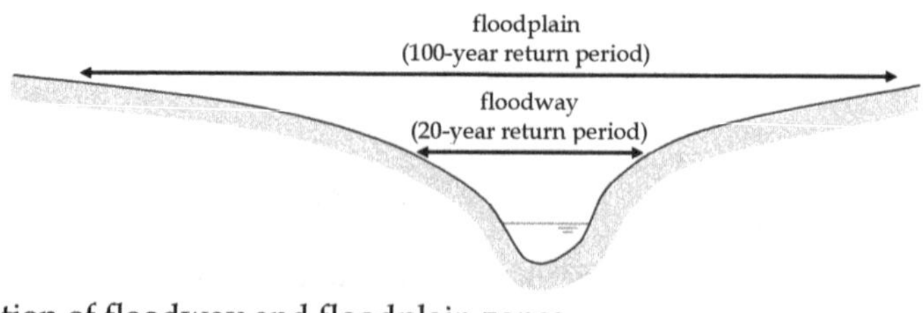

Fig. 13. Illustration of floodway and floodplain zones

Avoiding the encroachment of the floodway is extremely important and that is why building in this area is forbidden in many countries. These areas are more suitable for the development of public parks, which can act as multifunctional landscapes, or environmental conservation zones and can be managed in order to become greenways along the city.

In general, the occupation of the flood fringe is allowed, although sometimes with restrictions such as requiring the base floor level to be above the base flood (100-year return period, for instance) maximum water stage plus a certain safety margin freeboard or designing and constructing in accordance with flood-proofing building codes.

Flood zones can be represented as maps which should be considered as basic information for several urban planning and management activities. The development of these maps can be supported by GIS techniques and the resulting products should be available for free public access. A trend observed since the last decade is the development of combined packs joining hydrodynamic and hydrologic simulation programs with features provided by GIS software. Kraus (2000) shows some benefits concerning the use of GIS StreamPro to calculate and represent flood maps for the American National Flood Insurance Program (NFIP). According to Dodson & Li (2000) the time taken to produce flood maps with the aid of GIS based programs can be reduced in 66% compared to traditional approaches.

In the USA, the Federal Emergency Management Agency (FEMA) defines flood zones on its flood insurance rating map (FIRM). This is an example of a desirable integration between floodplain management and the NFIP.

Public authorities can also purchase and demolish properties in flood risk areas. In these cases, affected people and properties need relocation. This is a very common frame noticed in poor and developing countries. In Brazil, part of the money assigned to major drainage works is frequently destined to floodplain acquisitions and relocation of households.

3.2.2 Master Planning

Flood management master plans (FMMP) consist of a set of strategies, measures and policies arranged together in order to manage flood risk and guide the development of drainage systems.

One basic concept regarding master planning is that is should apply to the river basin as a whole. Additionally, this plan should be carried out integrated and harmonically with other urban planning and management instruments, regulations and related laws. In some countries, especially in wealthy ones or in cities with combined sewers systems, it is also frequent that part of the FMMP studies account for water pollution and soil erosion control. In the other hand, poor countries still face enormous difficulties regarding flood risk reduction and in these cases, generally, aspects related to water pollution and erosion control assume minor relevance.

Basically, a FMMP include different studies, data collection and programs, such as (adapted from Andjelkovic, 2001):

- the definition of goals and objectives that should be fulfilled in a foreseeable future;
- inventory of all drainage and flood control infrastructure;
- gathering hydrologic data regarding rain and river gages as well as past flood records;
- a diagnosis of flood problems and its causes;
- analysis of existing stormwater practices and its inadequacies;
- flood zoning studies in order to determine land use restriction;
- proposal of feasible structural and non-structural measures;

- design and cost estimate of proposed works and measures;
- benefit/cost analysis and comparative evaluation of alternative solutions;
- definition of drainage facilities design criteria;
- water pollution and soil erosion control program; etc.

3.2.3 Flood Forecasting and Warning

Early warnings can save lives and significantly reduce tangible and intangible losses due to natural hazards. In developed countries, the use of flood forecast and warning systems, such as those implemented for the Danube and the Mississippi river basins, represents one of the main trends in terms of non-structural flood control measures and has shown highly effective in reducing flood losses (Smith, 1996).

Some case studies authors claim that, in theory, flood damage reduction can reach up to two thirds of total losses. Actually, the reduction of economic losses effectively achieved through this kind of measure is about half of this estimate (Smith, 1996).

Flood forecasting in large basins is much simpler than in small ones, which are usually affected by flash floods. This is mainly due to the difficulties and uncertainty regarding the forecast of storms with short duration and concentrated in small areas.

One factor that substantially affects flood damage reduction is the warning lead time. Penning-Rowsell et al. (2003) developed curves relating flood warning lead time and flood damage reduction.

The expect benefits of a warning system depends not only on an efficient communication strategy to the people living in prone areas, but also rely on the level of preparedness of the affected community. The development of educational actions focusing on an increase of people awareness and preparedness can strengthen local community to face floods. This action can be carried out through public workshops and hearings, as well as using web communication or even on paper leaflets to be distributed (Andjelkovic, 2001). Emergency response teams can also take advantage of flood warning systems.

Another concern regarding the functioning of these systems relates to the uncertainty of the forecast. Fake alerts usually tend to reduce the population's reliability in the warning system and community coping with flood reduction strategies.

3.2.4 Flood Proofing

Flood proofing consists in the use of permanent, contingent or emergency techniques to prevent flood water from reaching buildings and its contents, as well as infrastructure facilities, or to minimise flood damage (Andjelkovic, 2001). Basically, the design of flood proofed constructions must consider floodwaters forces due to flooding depth, flow velocities and debris impact potential. There are several types of flood proofing techniques, as shown in figure 14. Some of the adjustments that may be necessary to ensure flood proof of a building are: anchoring it to withstand flotation, lateral movements and collapse; installation of watertight closures for door and windows; reinforcement of walls; installation of check valves to prevent entrance of stormwater or sewage through utilities; location of electrical, mechanical and other damageable equipment above expected flood level; floodwalls, small levees, berms or other kinds of barriers; among many other possible actions (FEMA, 1993).

Urban policies or floodplain regulations can require new constructions in the floodplain zone to comply with a flood proofing building code. Existing building can also be retrofitted in order to improve its flood protection level.

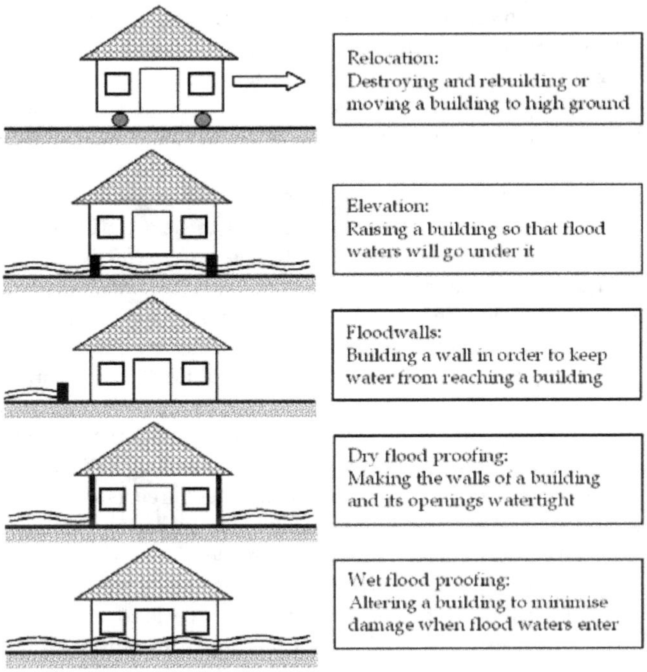

Fig.14. Examples of flood proofing measures (adapted from UNESCO, 1995)

3.2.5 Other Measures

Besides those non-structural measures previously listed, there are several other possibilities of application of this kind of measure. Environmental education activities and the establishment of a flood insurance program are other examples of non-structural flood risk management alternatives.

4. Urban Flood Models

Mathematical modelling of physical processes is a valuable tool to understand their systemic behaviour and the interactions among their individual components. However, practical solutions for mathematical models demand the introduction of a set of simplifications to be considered. Natural phenomena, because of their diversity, are generally not simple to model. Depending on the hypothesis considered, one given model may be suitable for certain situations, but may not be applied to other conditions. Although the choice of an adequate model may be difficult, models must be an active part of planning or design solutions, especially where the considered problem demands a systemic approach or when future scenarios must be analysed. The predictive capacity of a mathematical model is one of its most distinguishable characteristics to be valued (Cunge et al., 1980).

When dealing with floods, there are complex aspects related to spatial and temporal flow variations. At urbanised basins, topography and man-made landscapes interact to increase the diversity of possible flow patterns. Urban floods may become a difficult challenge, when drainage net fails and surcharged pipe flow occurs, jointly with open channel flow and flow over streets, composing a complex picture where hydraulic structures and typical structures of urban landscape interact to redefine a practical drainage net, not planned and not desired. This situation leads to great flooded areas with lots of losses of different kinds. Flood solutions must consider the whole system interactions, not transferring problems downstream nor combining undesirable effects. It is important to maintain track of what is happening in different parts of the watershed, in order to avoid peak combination of floods coming from different sub basins.

Integrated projects for urban flood control have to identify how to optimise benefits of different individual measures considered together, and these are difficult questions that can be treated with the aid of mathematical models. In this context, it is important to recognise that choosing an adequate model is the first task when dealing with systemic problems.

The basic needs associated to an adequate urban flood model may be resumed below:

- The correct identification and characterisation of the problem, in order to understand main causes of the process and to choose suitable simplification hypothesis for a sound modelling formulation;
- Sometimes, when designing drainage net, one-dimensional modelling can be applied, once it is expected that there will be no overflow for the design discharge adopted. Other times, even when overbank flow occurs, if the flooded area is confined alongside river course, it is possible to use one-dimensional model, extrapolating calculated channel water levels. However, when inundation of great areas leads to flow patterns dictated by topography, with little relation to channel flow, or when an urban area suffers from lack of adequate micro-drainage and flooding begins with overland flow accumulation, two-dimensional or pseudo two-dimensional models are more suitable;
- On the last case mentioned in the previous item, it is important to consider that the proposed model must be able to join drainage net with urban landscape, as it is possible that streets will act as channels, squares, parks, parking lots and buildings will act as undesired reservoirs, walls and roads will be barriers to the flow, at lower levels, but will become to act as weirs, when flooding levels rise;
- Considering the diversity of a urban drainage system, it is important that the model can be able to simulate different hydraulic structures, as weir, orifices, pumps, flap gates, etc.

4.1 Hydrologic Aspects

Hydrology studies are necessary in order to determine peak flow rates or the design hydrograph, depending on the type of study carried out. The focus of hydrologic flood modelling is to represent rainfall-runoff transformation. It is also often necessary to determine a design rainfall, as it is the basic input considered in this process.

One of the main issues with which engineers must deal in order to develop urban flood studies is the definition of the hydrologic approach to be used. Choosing a suitable methodology depends on physical characteristics of the catchment and also on the available data and the study goals. Ponce (1989) proposes a simplified scheme that presents adequate approaches according to basin size, as seen in figure 15. As shown in this figure, the rational method meets the requirements needed in small catchment applications (usually limited to 2,5km^2 areas), unit hydrograph techniques suits better midsize watersheds and routing methodologies are suitable for large basins simulation.

The representation of the hydrologic cycle or part of it is the basis of engineering hydrology methods. Due mostly to the time scale of urban floods, some components of this cycle can be neglected. Evaporation, transpiration and groundwater flows variations are slow processes that have no significant effect on flood hydrographs. Therefore, the most important phenomena are precipitation, infiltration, vegetal interception and depression storage (which are usually considered combined and denoted as initial losses or abstraction) and surface runoff.

Method or approach	Catchment scale: Small	Midsize	Large
Rational method	Usually	Not applicable	Not applicable
Unit hydrograph	Not applicable	Usually	Not applicable
Routing tehcniques	Not applicable	Not applicable	Usually

Fig.15. Suitable methodological approaches according to basin size (Ponce, 1989)

The representation of the hydrologic cycle or part of it is the basis of engineering hydrology methods. Due mostly to the time scale of urban floods, some components of this cycle can be neglected. Evaporation, transpiration and groundwater flows variations are slow processes that have no significant effect on urban flood. Therefore, the most important phenomena are precipitation, infiltration, vegetal interception and depression storage (which are usually considered combined and denoted as initial losses or abstraction) and surface runoff.

Hydrologic models consist of a set of mathematical equations arranged in order to describe relevant phases of the hydrologic cycle and can be classified according to different features. Some of the main types of models are:

- physical or mathematical – the first depends on a physical representation of the prototype and, in practical hydrology, is almost never used, while the second is based on mathematical equations and constitutes more common tools;
- theoretical, conceptual or empirical – a theoretical model is based on general governing physical laws, an empirical model is based on equations using parameters determined from data analysis and conceptual models are based on either theoretical and empirical equations in order to try to represent system behaviour;
- single-event or continuous streamflow simulation– the model can represent the catchment hydrologic response for only a single storm event or determine streamflow regime in a continuous basis;
- lumped or distributed – lumped models can describe rainfall and flow rate temporal variations but cannot represent spatial variations, while distributed models are capable of describing both of them (Ponce, 1989);
- deterministic or stochastic – the difference between these kind of models is that the response of a determinist model to a given input data is always the same, while the relation between input and output in a stochastic model depends on random properties of the time series.

The continuous development of computers over the last decades has been stimulating the use of mathematical models. This happens due to the ever increasing availability of computers and the progress of computer sciences and processing capability.
The most common type of model used in flood hydrology applications is the simple, single-event, rainfall-runoff simulation model. The primary interest of these models is the determination of the flood hydrograph.
The basic set of information needed to develop a flood hydrology study is:

- rainfall and streamflow data;
- rain gages intensity-duration-frequency equations;
- rainfall depth-area-duration curve for the region (not applicable for small catchments);
- topographic mapping of the catchment;
- land use mapping;
- soil types mapping;
- unit hydrograph (if available, otherwise it is possible to use synthetic hydrographs).

Applications based on distributed models are more complex and usually need a large amount of data for its calibration. However, frequently there is no availability of the required data, as its collection is expensive and difficult. This kind of models also face scaling challenges, due to the difference of field measurements, which are representative of a point or a local scale, and the computational grid used to represent hydrologic processes (DeVries & Hromadka, 1993).
The following paragraphs present a simple description of a typical hydrologic design sequence for flood peak calculation in midsize basins using synthetic precipitation and hydrographs. For a broader discussion on the available methods, their characteristics and its limitations it is suggested that the reader refer to specific books such as Linsley et al. (1984); Ponce (1989); Hromadka II et al. (1987) ; Urbonas & Roesner (1993); among others.
The time of concentration is usually defined as the period necessary for the runoff produced in the most remote point of the catchment to reach a given point or cross-section. It is frequent to consider rainfall critical duration as equal to the catchment time of concentration. This hypothesis is suitable for small watersheds, reasonable for midsize catchments, but not applicable to large basin. In theory, the time of concentration is composed by two different parts: time to equilibrium and time of travel. There are several equations developed to calculate the time of concentration in catchment with different characteristics. Some of it focus mainly in overland flow representation (which is associated to time to equilibrium), while others are concerned mostly with the account of the time of travel. In small catchments the time to equilibrium is the preponderant parcel and, in the other hand, the time of travel is the most important in large basins. Hence, it is important to know the applicability limits of each formulation aiming to choose a suitable approach.
In order to determine a design storm it is necessary to define a return period associated with this event. High return periods lead to a lower risk of flooding and to higher costs of the necessary flood control works.
It is common to calculate the design storm using an intensity-duration-frequency curve, which refers to a specific rain gage. The frequency is related to the storm return period. Higher return periods implies in higher precipitation depths and intensities. Rainfall duration affects this curve in a different way. Higher storm intensities are achieved with lower duration, while total rainfall depth increases with duration.

There are some methods that can be used to determine an average precipitation over an area, such as: Thiessen polygons method; isohyetal method; and average rainfall method (as kwon as arithmetic method). As catchment area grows, it becomes necessary to correct rainfall through a depth-area-duration adjustment curve.
Rainfall can be represented in three different ways: constant in both space and time; constant in space but varying in time; and, varying in both space and time. The first approach is suitable for small catchments, while the second and third hypotheses are adequate to midsize and large basins, respectively (Ponce, 1989). A distributed model is required in order to represent spatial variations.
Once the design rainfall is defined, the next step is to calculate runoff depth, or precipitation excess. There are many methods that can be used for this purpose, such as: the rational method; the Soil Conservation Service (SCS) method; the use of phi-index method; the use of potential infiltration curves, such as the Horton formula, for instance; among others. Some of these methods are more suitable for small watershed, while others are more indicated to larger catchments.
Finally, the last step to determine a flood hydrograph can be carried out with the aid of synthetic unit hydrograph methods. These methods assume that the catchment behaviour is linear, which implies that if the basin response for a unit rainfall is know, one can determine its response for any rainfall. There are several synthetic unit hydrographs methods such as, in example, the SCS method or the Snyder method.

4.2 Hydrodynamic Aspects

Hydrodynamic aspects of urban flood modelling encompass various typical aspects of general flood modelling. The hydrodynamic model must use the mass conservation law and hydraulic and hydrodynamics relations as the core engine. The Saint-Venant equations are usually used to represent flow conditions in the main channel net. This system of equations may appear in a one-dimensional form, a two-dimensional form or in a pseudo two-dimensional form, where a spatial region is divided into an integrated mesh of cells, linked by one-dimensional equations, although composing a two dimensional flow net.
An alternative way to represent flow mass balance, appropriated to a cell flow model representation, considers that the water level variation in a cell i, at a time interval t, is given by the continuity equation applied for that cell as stated in equation (1).

$$A_{S_i}\frac{dZ_i}{dt} = P_i + \sum_k Q_{i,k} \tag{1}$$

Where: $Q_{i,k}$ is discharge between neighbours cells i and k; Z_i is the water surface level at the centre of the cell i; A_{S_i} is the water surface area for the cell i; P_i is the discharge related to the rainfall over the cell; and t is a independent variable related to time.
River and channel flows, as well as flow over the streets, may be represented by the Saint-Venant dynamic equation. Taking into account a rectangular cross section and a fixed bottom result in equation (2) (Cunge et al., 1980).

$$\frac{1}{A_{i,k}}\frac{\partial Q_{i,k}}{\partial t} - \frac{QB_{i,k}}{A_{i,k}^{\ 2}}\frac{\partial Z}{\partial t} + g\frac{\partial Z}{\partial x} + gS_f = 0 \tag{2}$$

Where: $B_{i,k}$ is the surface flow width between cells i and k; $A_{i,k}$ is the wetted flow cross-section area between cells i and k; S_f is the energy line slope; $R_{i,k}$ is the hydraulic radius of the flow cross-section between cells i and k; n is Manning's roughness coefficient; and x, t are independent space and time variables.

The diversity involved in the detailed representation of the urban watershed may require various other hydraulic laws, in order to represent different types of flow.

One question that must be emphasised is that the whole basin must be represented. This consideration allows a systemic modelling with a comprehensive approach that may simulate the integrated consequences of acting over different parts of the basin, inside and outside drainage net. This is what makes a model really useful, especially in flood control planning. Representing the whole basin, however, can reveal a very difficult task, depending on the scale of interest. When parts of a watershed do not present any special interest, it is possible to substitute these parts by boundary conditions that concentrate the effects of the outer parts of the basin at the interface between modelled area and outside areas. Boundary conditions may represent, discharge series, water level series or discharge vs. water levels relations. Figure 16 pictures a region schematically modelled, showing an arrangement of cells, where mathematical equations are applied and boundary conditions substitute parts of the basin not modelled. In this example, upstream boundary conditions represent the discharges of upper basin reaches, while downstream condition represents water levels showing tidal influence at a hypothetical bay.

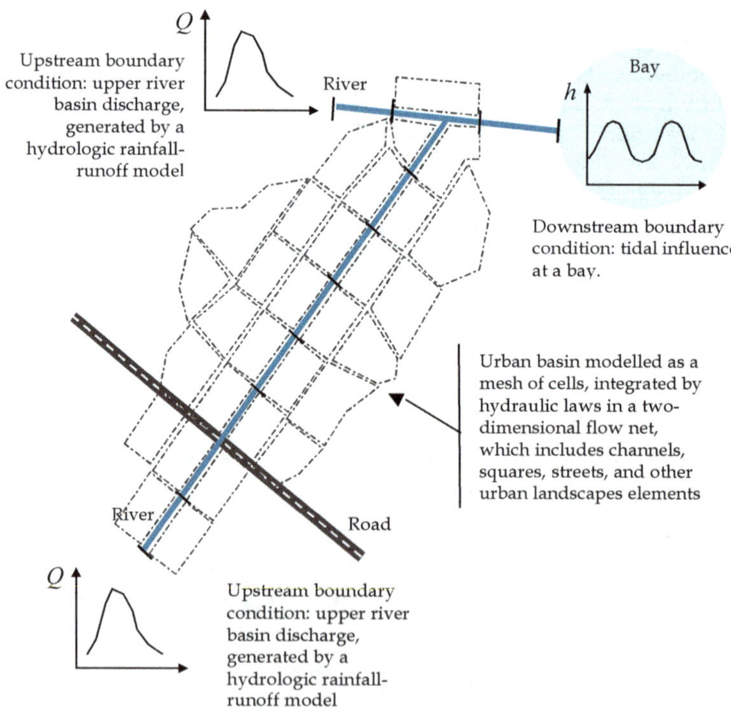

Fig. 16. Hypothetical mathematical modelling of an urban basin

4.3 Illustration of a Set of Typical Urban Flood Model

Urban flood modelling is increasing in interest, once urban floods appear as one of the most frequent, serious and costly problems that cities must face. Many models, with different characteristics, may be cited. In order to illustrate the discussion held in this chapter, some of the most common programs used for urban flood simulation will be mentioned in the following paragraphs.

MIKE FLOOD is a comprehensive modelling package, developed by DHI, covering the major aspects of flood modelling. MIKE FLOOD integrates flood plains, streets, rivers and sewer/storm water systems into one package. In order to achieve this objective, MIKE FLOOD join three widely used hydrodynamic models namely MIKE 21, MIKE 11 and MIKE URBAN into one package. This way, a 1D model and a 2D model are coupled with a sewer model, enabling analysis of flooding and assessment of the consequences of planned solutions The philosophy adopted allows an appropriate spatial resolution, so that. pipes and narrow rivers are modelled using one-dimensional solvers whereas the overland flow is modelled using two spatial dimensions. Some characteristics of MIKE FLOOD are: coupled one and two-dimensional flow, integration of hydraulic structures in 2D grids, effective mass conserving flooding/drying routine, accurate and physically based simulation of flow splits (DHI, 2008).

The United States Environmental Protection Agency (EPA) developed Storm Water Management Model (SWMM), which is a dynamic rainfall-runoff simulation model used for single event or long-term (continuous) simulation of runoff quantity and quality from primarily urban areas. The runoff component of SWMM operates on a collection of subcatchment areas that receive precipitation and generate runoff and pollutant loads. The routing portion of SWMM transports this runoff through a system of pipes, channels, storage/treatment devices, pumps, among others. SWMM tracks the quantity and quality of runoff generated within each subcatchment, and the flow rate, flow depth, and quality of water in each pipe and channel during a simulation period comprised of multiple time steps (Rossman, 2008).

SWMM was first developed in 1971, and has undergone several major upgrades since then, being used for planning, analysis and design related to stormwater runoff, combined sewers, sanitary sewers, and other drainage systems in urban areas, with many applications in non-urban areas as well. The current edition is SWMM 5.

The Hydrologic Modelling System (HEC-HMS), developed by US Army Corps of Engineers (USACE) is designed to simulate the precipitation-runoff processes of dendritic watershed systems. It is designed to be useful in a wide range of geographic areas, including large river basin water supply and flood hydrology, and small urban or natural watershed runoff. Hydrographs calculated by the program are used directly or in conjunction with other software for studies regarding water availability, urban drainage, flow forecasting, future urbanisation impact, reservoir spillway design, flood damage reduction, floodplain regulation, and systems operation.

The program is a generalised modelling system capable of representing many different watersheds. A model constructed for one watershed considers separation of the hydrologic cycle into manageable pieces and the definition of boundaries around this watershed, in the area of interest. In most cases, several model choices are available for representing each kind of problem (Scharffengerg & Fleming, 2008).

4.4 MODCEL – An Overview

MODCEL (Mascarenhas et al., 2005) is an urban flood model, which integrates a hydrologic model, applied to each cell in the modelled area, with a hydrodynamic looped model, in a spatial representation that links surface flow, channel flow and underground pipe flow, This arrangement can be interpreted as a hydrologic-hydraulic pseudo 3D-model, although all mathematical relations written for the model are one-dimensional. Pseudo 3D representation may be materialised by a hydraulic link taken vertically to communicate two different layers of flow: a superficial one, corresponding to free surface channels and flooded areas; and a subterranean one, related to free surface or surcharged flow in galleries The construction of MODCEL, based on the concept of flow cells (Zanobetti et al., 1970) intended to provide an alternative tool for integrated urban flood solution design and research. The representation of the urban surface by cells, acting as homogeneous compartments, in which it is performed rainfall run-off transformation, integrating all the basin area, and making it interact through cell links, using various hydraulic laws, goes towards the goals to be achieved by the mathematical modelling of urban floods, as discussed in the previous sections. Different types of cells and links give versatility to the model. Figure 17 shows a catchment's profile, where it is possible to see a cell division and the interaction between cells.

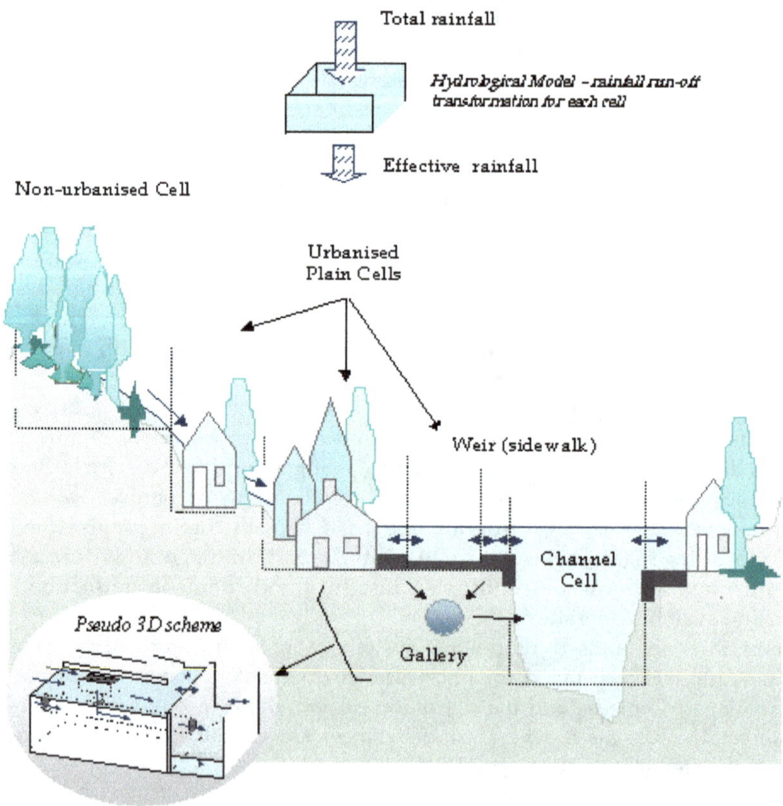

Fig. 17. Schematic vertical plane cut in an urban basin showing a cell model representation

The cells, solely as units or taken in pre-arranged sets, are capable to represent the watershed scenery, composing more complex structures. The definition of a set of varied flow type links, which represent different hydraulic laws, allows the simulation of several flow patterns that can occur in urban areas. Therefore, the task related to the topographic and hydraulic modelling depends on a pre-defined set of cell types and possible links between cells.

The pre-defined set of cell types considered in MODCEL is listed below:

- River or channel cells - are used to model the main free open channel drainage net, in which the cross section is taken as rectangular and may be simple or compound;
- Underground gallery cells - act as complements to the drainage net;
- Urbanised surface cells - are used to represent free surface flow on urban floodplains, as well as for storage areas linked to each other by streets. Alternatively, these cells can represent even slope areas, with little storage capacity. In this case, they are designated to receive and transport the rainfall water to the lower modelled areas. Urbanised plain cells can also simulate a broad crested weir, which conduct water spilled from a river to its neighbour streets. These kinds of cells present a gradation level degree, assuming a certain pre-defined storage pattern, as shown in figure 18;
- Natural (non-urbanised) surface cells - these cells are similar to the preceding case, however having prismatic shape without considering any kind of urbanisation;
- Reservoir cells - used to simulate water storage in a temporary reservoir, represented by an elevation versus surface area curve.

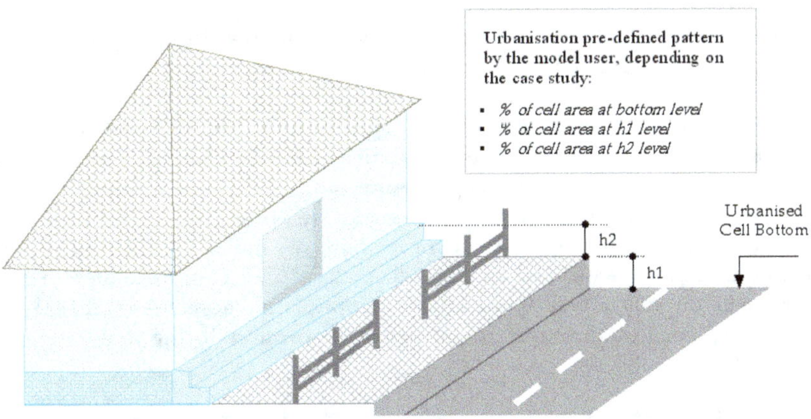

Fig. 18. Urbanisation storage pattern representation

Typical hydraulic links between cells can be summarised as shown below (Miguez, 2001; Mascarenhas et al., 2005):

- River/channel link - this type of link is related to river and channel flows. It may eventually also be applied to flow over the streets. More specifically, it corresponds to the free surface flow represented by the Saint-Venant dynamic equation;
- Surface flow link - this link corresponds to the free surface flow without inertia terms, as presented in Zanobetti et al. (1970);
- Gallery link - this link represents free surface flow in storm sewers, as well as surcharged flow conditions. Free surface flow is modelled the same way as in surface flow links, using simplified Saint-Venant dynamic equation. On the other hand, when galleries become

drowned, pressure flow conditions are given by energy conservation law; therefore, using Bernoulli equation;

- Inlet gallery link/Outlet gallery link - computed flow conditions define if the inlet/outlet is drowned or not, also considering the possible occurrence of local head losses;
- Broad crested weir link - this link represents the flow over broad-crested weirs. It is used, mainly, to represent the flow between a river and its margins;
- Orifice link - this link represents the classic formula for flow through orifices;
- Street inlet link - this link promotes the interface between surface and gallery cells. When not drowned, this link acts as a weir conveying flow from streets to galleries. When drowned, this link considers flow occurring through a certain number of orifices associated to the street inlets;
- Reservoir link - this link combines an orifice, as the outlet discharge of a reservoir, with a weir, that can enter or not in charge, depending on reservoir operation;
- Stage-discharge curve link - this link corresponds to special structures calibrated at physically reduced scales in laboratory and basically relates a discharge with a water level, in a particular equation;
- Pumping link - this link allows to pump discharges from a cell to another, departing from a starting pre-defined operation level;
- Flap gate link - this link simulates flows occurring in the direction allowed by the flap gate opening, and can be often found in regions protected by polders.

4.5 Acari River Mathematical Modelling - A Case Study in a Poor Region of RJ/Brazil

The basin of the river Acari has a drainage area of about $107km^2$, composed by densely populated neighbourhoods of the city and containing several important streets, avenues and highways. This region, however, is one of the most poor of the city and there are various informal communities established there, especially near river banks. The main river itself shows signs of heavy environmental degradation, with solid waste disposal, garbage and sediments appearing in several reaches. Flooding is one of the critical problems of the basin as well. There are inundation records of more than one meter in different places. At the critical points, there are records of almost two meters. City Hall estimates that floods on Acari river basin directly affect about 20,000 people, and more than 150,000 people are affected indirectly, because of urban infrastructure disruption during inundation. Figure 19 shows some of these problems.

Fig. 19. Scenes of Acari river basin

The solution for Acari River basin floods poses a difficult problem, combining critical flooding levels, social pressures, lack of appropriated infrastructure, sea and tidal influence. The first attempt to treat this problem, as proposed by Rio de Janeiro City Hall, referred to the traditional approach of canalisation. This design concept arose because of several detected river bed obstructions and river banks occupation, facts that suggested the need of improving conveyance. However, this proposition would probably not be able to solve the problem by itself. Tide at the outlet of the basin limit the discharge capacity and large flooded areas spread around the basin show that simple canalisation would transfer the problem to lower areas, increasing flood magnitude at these parts of the basin.
Facing this problem, Rio de Janeiro City Hall and Federal University of Rio de Janeiro joined efforts in the search of a systemic solution, balancing conveyance and storage approaches. The basin, showed in figure 20, was modelled using MODCEL. An example of the cell division, is provided in figure 21.

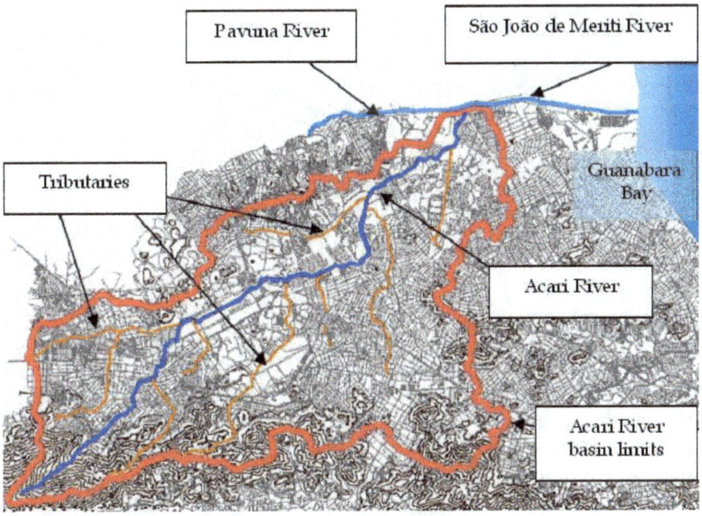

Fig. 20. Plain view of Acari river basin

Fig. 21. Detail of the cell division for Acari River Basin Modelling

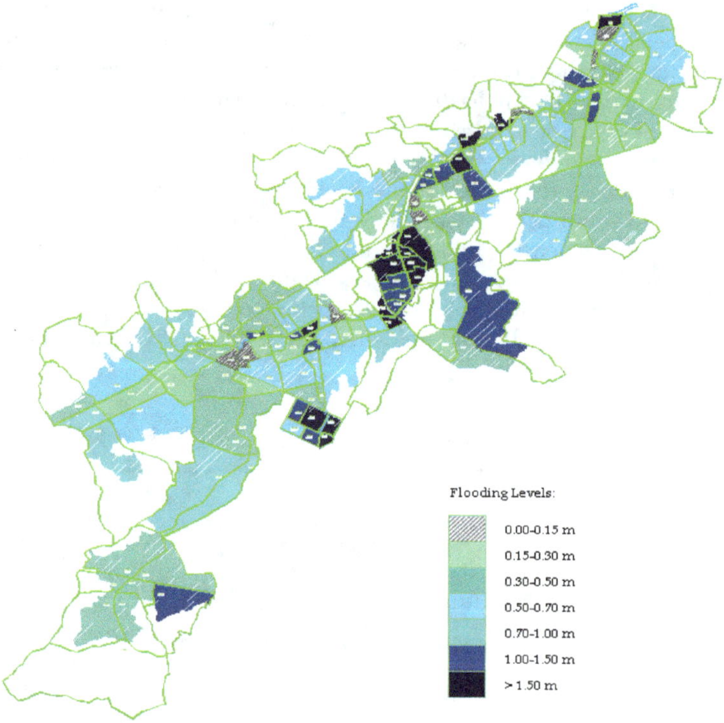

Fig. 22. Flood map for present situation

After analysing flood patters and making a diagnosis of the flooding present situation, whose flood map is seen in figure 22, a set of complementary and integrated measures was proposed as a result of prospecting scenarios generated by the model:

- canalisation was not considered necessary in a large scale, although it should be useful and recommended for specific reaches;
- it was necessary to propose an dredging of medium and low reaches of the river, in order to deal with river bed sedimentation and local obstructions, specially near bridges pillars;
- low bridge beams, working as local barriers to flood flow, must be remodelled (one of the bridges, at Luis Coutinho Cavalcanti street was considered very critical);
- the original storage capacity of the basin needs to be, at least, partially restored. In this way, a set of reservoirs was proposed, with two major reservoirs in important tributaries of Acari River. Other measures included one detention basin proposed in the left margin of the river, near a military area, and a slum area, on the right river margin, was proposed to turn into a park and to work as a multifunctional landscape, damping high discharges;
- people living in very critical areas, in the flood plains, needs to be relocated to safer areas;
- flood problems could be reduced, but there would be areas still strongly affected. It is important to understand that only a long-term work could produce better results. Sustainability needs a larger range of actions. Environmental recovery and investment in general urban infrastructure are necessary to revert the situation. Education and economic development complete the puzzle to construct the desired solution for the problem.

After considering this set of interventions, comparing flood levels at 18 control points, there was an average reduction of 30%. The higher water level reduction result showed inundation diminished by 76% (from 1.31m to 0.31m).

4.6 Use of soccer fields as complementary areas of a temporary storage pond in a poor community

This second case study refers to a region of Rio de Janeiro State (RJ) known as Baixada Fluminense, located at the metropolitan region of Rio de Janeiro City and occupied mostly by low-income families. This region is also characterised by low level lands naturally subject to floods caused by Iguaçu and Sarapuí rivers. Dikes have been built to prevent the flooding of this region, and as a consequence, polder areas were created. The typical arrange of these polders consists of a stormwater temporary storage pond which receives the major drainage channels and is connected to Iguaçu or Sarapuí rivers through flap gates. The use of flap gates to allow discharge of these polders has the advantage that this kind of structure is passive, robust and requires no operation. The disadvantage is that the discharge can only take place during low tides and these periods can sometimes be delayed due to the routing of floods in the Iguaçu and Sarapuí rivers and adverse climatic conditions. Pump stations could overcome these limitations, but the use of this kind of solution in such case can be considered inappropriate due to the lack of security of the facilities and high operation and maintenance costs. As a result, in order to prevent the water from rising up to a certain level that could cause uncontrolled flood of the surrounding area and consequent failure of other elements of the drainage system, a greater temporary storage volume is required.

Polder Alberto de Oliveira, which receives drainage of part of São João de Meriti and Duque de Caxias municipalities (RJ), is taken in this case study as an example of what is occurring with other polder areas at Baixada Fluminense region. Regular and irregular buildings have been occupying a portion of almost 80% of polder original area designed to work as stormwater temporary storage pond (COPPETEC, 2003). Visiting this community, it can be observed that one of the measures developed by local population, in order to prevent flood losses, was building their homes over 1.0 to 1.5 meter tall pillars. Urbanisation of the catchments also aggravates the problem, as the runoff production got higher than that estimated by the time the original pond was designed. These two factors caused the flood risk of the region to rise considerably. Recent storms and the extension of flooding areas caused a lot of public pressure over the municipalities and state governments. The response of the authorities was the creation of a program to reduce the flood risk in this area. So forth, studies have been carried out in order to determine which interventions are needed to maintain the water inside the pond, considering a maximum water level that could cause no flood hazard to the surrounding community. MODCEL (Miguez, 2001) was used to simulate the flood at the polder area and at the Sarapuí River. A 20-years return period storm was set for the polder area and a 10-year return period storm was used for the Sarapuí river basin.

The results of the mathematical simulation showed that three combined possibilities could reduce water level in the storage pond area to the desirable level (COPPETEC, 2003): a) double the number of flap gates; b) set a 8m3/s pump station close to the remaining storage area; c) reallocate part of the population that occupies the original temporary pond area. Due to the already mentioned problems concerning pump facilities this alternative has been abandoned. One demand of state authorities was the reduction of the number of families in need of reallocation. The final scenery proposed considered an increase of the number of

flap gates (60% more flow capacity) and the lowering of the ground level of two areas close to the remaining storage pond. Few families occupy one of these areas and several soccer fields occupy the other. Figure 23 shows the cell division of the region and these areas.
An interesting aspect about the behaviour of local communities in Brazil is that it is very hard to prevent the occupation of free spaces close to poor communities, but soccer field areas are almost always respected, as there is a public perception that these areas serve as leisure and sport facilities for the community. Part of the strategy was setting a multifunctional landscape at the soccer fields' area, so that it could assume a new function, flood control. The proposal was lowering this area to a ground level higher than the other new storage area which is being added to the remaining pond, so that this complementary storage volume gets used only in case of more intense storms, allowing its sportive function at most of the time. The set of measures presented in the final scenery are currently under construction.

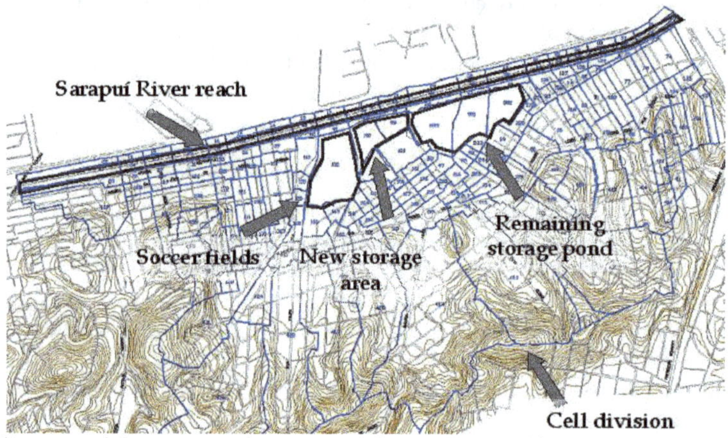

Fig.23. Cell division of the region of interest and new areas added to the remaining pond

5. Concluding Remarks

Flood control is one of the major questions with which urban planners must deal nowadays. According to Freeman (1999), 60% of human life losses and 30% of economic losses caused by natural disasters are due to floods. Besides, urban floods involve several different aspects in a mosaic involving climatic, technical, social, economic and environmental issues.
Technically, the urban flood problem must be understood in both spatial and temporal dimensions. In this context, city landscape diversity aggregates one more difficulty, generating a complex flow pattern.
Optimal Engineering solutions are not always possible to be achieved because of social or political and institutional constraints. However, in order to have the best possible solution, it is necessary to provide integrated, sound and efficient design alternatives.
In this context, mathematical modelling can provide an important tool to aid in the design process. Models allow the recognition of flood patterns and urban drainage behaviour, enabling the capability of creating different future scenarios of urban growth and proposed design concepts to deal with the problem. Stormwater in cities is a matter to be managed linked with land use planning.

Classic site-specific planning needs to be replaced by a watershed oriented planning. Local and isolated solutions tend to transfer flood problems. The traditional canalisation approach, improving conveyance and focusing the consequences of floods, cannot face alone the flooding problem. New approaches focus on storage and infiltration measures, as well as on preventive actions, complementing the traditional ones. Therefore, the concepts applied to stormwater drainage design have been changing a lot in the past decades, pointing to a systemic approach. Structural measures, of different kinds, are being proposed to reorganise flow patterns and partially recover hydrologic conditions previous to urbanisation, while non-structural measures aim to provide rational coexistence with floods. All these changes along time and the state of art evolution detach the challenge with which cities are being faced: to find a sustainable path to equilibrate city growing with a harmonic built environment for their communities.

6. References

Andjelkovic, I. (2001). *Guidelines on Non-structural Measures in Urban Flood Management.* Technical Documents in Hydrology. UNESCO, Paris

AMEC (2001). *Earth and Environmental Center for Watershed Protection.* Georgia Stormwater Management Manual, vol.2: Technical Handbook. Atlanta, USA

Arizona (2003). *Harvesting Rainwater for Landscape Use.* [on line], Internet url: http://ag.arizona.edu/pubs/water/az1052/harvest.html

Butler, D. & Davies, J.W. (2000). *Urban Drainage,* ISBN 0419223401, London, England

Coffman, L.S., Cheng, M., Weinstein, N. & Clar, M. (1998). *Low-Impact Development Hydrologic Analysis and Design.* In: Proceedings of the 25th Annual Conference on Water Resources Planning and Management, Chicago-Illinois, USA, p. 1-8

COPPETEC (2003) *Mathematical Modelling of Alberto de Oliveira Polder.* Final Technical Report, PEC 3850, Brazil *(in Portuguese)*

COPPETEC (2004) *Mathematical Model of Urban Floods, using Flow Cell Concepts, as a Management Tool for Integrated Flood Control Design Projects.* Final Technical Report, PEC 4221-CT-Hidro/GBH nº 520093/2003-8, Brazil *(in Portuguese)*

COPPETEC (2007) *Environmental Recovery and Integrated Flood Control Design Projects for Guerenguê River Basin at Rio de Janeiro City.* Final Technical Report, POLI-8498, Brazil *(in Portuguese)*

Cunge, J.A., Holly Jr., F.M. & Verwey, A. (1980). *Practical Aspects of Computational River Hydraulics.* Pitman Ad. Publishing Program, ISBN 0273084429, London, England

DeVries, J.J. & Hromadka, T.V. (1993). Computer Models For Surface Water. In: *Handbook of Hydrology* (Ed. Maidment, D. R.). McGraw Hill

DHI (2008). *www.dhigroup.com/software/waterresources/MIKEFLOOD.aspx,* access in May 23

Dodson, R.D. & Li, X. (2000). The Accuracy and Efficiency of GIS-Based Floodplain Determinations. In: *Hydrologic and Hydraulic Modelling Support with Geographic Information Systems* (Ed. Maidment, D. & Djokic, D.). ESRI Press, Redland, USA

FEMA (1993). *Non-Residential Floodproofing - Requirements and Certification for Buildings Located in Special Flood Hazard Areas in Accordance with the National Flood Insurance Program.* Federal Emergency Management Agency, Washington, USA

Freeman, P. (1999). *Gambling on Global Catastrophe.* Urban Age, Vol. 7, n°1, Summer, p 18-19, Washington, DC, USA

Hunter, M.R. (1994). *Identification of Problems, Solutions and Cost Savings for Maintenance of Drainage Ways*. In: Urban Drainage Rehabilitation Programs and Techniques. American Society of Civil Engineering, p. 194-208, New York, USA

Hromadka II, T.V., McCuen, R.H. & Yen, C. (1987). *Computational Hydrology in Flood Control Design and Planning*. Lighthouse Publications. California

Kraus, R.A. (2000). Floodplain Determination Using ArcView GIS and HEC-RAS. In: *Hydrologic and Hydraulic Modelling Support with Geographic Information Systems* (Ed. Maidment, D. & Djokic, D.). ESRI Press, Redland, USA

Leopold, L.B. (1968). *Hydrology for Urban Planning - A Guide Book on the Hydrologic Effects on Urban Land Use*. USGS circ. 554, USA

Linsley, R.K., Kohler, M.A. & Paulhus, J.L.H. (1984). *Hydrology for Engineers*. Third Edition. McGraw Hill. Singapore

Macaitis, W.A. (1994), *Urban Drainage Rehabilitation Programs and Techniques*. American Society of Civil Engineering, ISBN 0784400385, New York, USA

Miguez, M.G. (2001). *Mathematical Flow Cell Model for Urban Basins*. D.Sc. Thesis, COPPE/UFRJ, Rio de Janeiro, Brazil. *(in Portuguese)*

Miguez, M.G., Mascarenhas, F.C.B. & Magalhães, L.P.C. (2007). *Multifunctional Landscapes For Urban Flood Control In Developing Countries*. Sustainable Development and Planning, Volume 2, Issue 2, WIT Press., Southampton, England and Boston, USA

Mascarenhas, F.C.B., Toda, K., Miguez, M.G. & Inoue, K. (2005) *Flood Risk Simulation*. WIT PRESS, ISBN 1853127515, Southampton, England and Boston, USA

Niemczynowicz, J. (1999). *Urban Hydrology and Water Management–Present and Future Challenges*. Urban Water, Volume 1, Issue 1, March, p.1-14, Elsevier, Netherlands

Penning-Rowsell, E.C., Johnson, C., Tunstall S., et al. (2003). *The Benefits of Flood and Costal Defence: Techniques and Data for 2003*. Flood Hazard Res. Centre, Middlesex Univ.

Ponce, V.M. (1989). *Engineering Hydrology*. Prentice-Hall. New Jersey

Rossman, L.A. (2008). *Storm Water Management Model 5.0 User's Manual*. United States Environmental Protection Agency, Cincinnati, OH, USA

Scharffengerg, W.A. & Fleming, M.J. (2008). *Hydrologic Modelling Systems HEC-HMS User's Manual*, US Army Corps of Engineers, Davis, CA, USA

SEMADS, (2001). *Floods in Rio de Janeiro*. Planágua Project SEMADS/GTZ, vol. 8, Rio de Janeiro, Brazil. *(in Portuguese)*

Simons, D.B. *et al.* (1977). *Flood flows, Stages and Damages*. Fort Collins: Colorado State University, USA

Smith, K. (1996). *Environmental Hazards, Assessing Risk and Reducing Disaster*. Routledge, London

UNESCO (1995). "Fighting Floods in Cities"; Project: Training Material for Disaster Reduction; Delft, Holland

Urbonas, B.R. & Roesner, L.A. (1993). Hydrologic Design for Urban Drainage and Flood Control. In: *Handbook of Hydrology* (Ed. Maidment, D. R.). McGraw Hill

Urbonas, B.R. & Stahre, P. (1993). *Stormwater Best Management Practices and Detention*, Prentice Hall, Englewood Cliffs, New Jersey, USA

Woodworth Jr., J.W. (2002). *Out of the Gutter, Reducing Polluted Runoff in the District of Columbia, USA*

Zanobetti, D., Lorgeré, H., Preissman, A. & Cunge, J.A. (1970). *Mekong Delta Mathematical Program Construction*. Journal of Waterways and Harbours Division 96, p. 181-199

5

Urban Engineering: Concepts and Challenges

Alex Abiko
University of São Paulo - Escola Politécnica
Alex.abiko@poli.usp.br
Brazil

1. Introduction

The purpose of this chapter is to explain the concepts of urban engineering and to highlight some of the challenges faced by this discipline. The overall idea is to describe how urban engineering relates to other areas of engineering expertise, particularly within the context of civil engineering. To do this we have drawn mainly on our own professional and academic experience, fleshed out by an examination of the relevant literature available both in Brazil and further afield. At the outset it should be said that most of our observations focus on the city of São Paulo where our present professional concerns lie. However, in future works we hope to extend our approach beyond the confines of São Paulo in an effort to broaden and improve our understanding of the concepts underlying urban engineering as a necessary prelude to enable us to supply useful guidance for researchers, experts and students keen to work alongside the engineering professionals currently employed in our cities.

2. Urban engineering in São Paulo

2.1 Background

The first topographical survey of the city of São Paulo was completed in 1792. According to Toledo (1983) the survey was in effect the first 'master plan' for the city. In addition to being a straightforward survey it also provided certain guidelines as to how the city should deal with its future expansion from small village to larger urban center.

The above survey was carried out by Portuguese military engineers, cartographers and astronomers belonging to the Royal Corps of Engineers, who were also engaged in overseeing a variety of public works such as the building of hospitals, the laying down of water facilities and paved streets, as well as constructing barracks and other military-type installations.

It is perhaps worth recalling that, prior to the late 18th century, so-called public works such as the construction of bridges and the paving of roads and streets tended to be undertaken by ordinary people using makeshift building techniques and perishable materials such as mud reinforced with straw (adobe). The Portuguese military engineers introduced a series of new techniques, employing more durable materials such as stone and lime (infinitely more suited to large-scale works).

Military-trained engineers played an important role in the development of the city of São Paulo and its hinterland, moving on from mapping and surveying the then "province" to undertaking topographical surveys of the expanding urban area, designing roads and railways and being closely involved in the construction of bridges, fortifications and public buildings in general (Simões Jr., 1990).

The growing importance of these activities, which expanded in tandem with the population upsurge in the interior of the state of São Paulo as a result of the coffee boom, pointed to an urgent need to train more engineers. The latter began to be referred to, around this time, as "civil engineers" given that the majority of the public works required were increasingly of a non-military nature.

The *Escola Politécnica* of São Paulo was established in 1893. This ran courses in civil, industrial and agricultural engineering as well as a supplementary course in mechanics. One year after its establishment the *Escola* was also able to offer courses in architecture and was entitled to award formal qualifications in accountancy, surveying and machinery operation for students who managed to complete only part of its engineering courses (Santos, 1985).

The first School of Engineering in Brazil to provide exclusively a course in civil engineering was the *Escola Politécnica* of Rio de Janeiro, established in 1874. The *Escola* originated in 1792 with the creation of the Royal Academy of Artillery, Fortifications and Design in Rio de Janeiro, which later (in 1810) became known as the Royal Military Academy. The Academy was in the event staffed by the director and most of the members of the teaching corpus who had previously worked at the Portuguese Royal Naval Academy, having arrived in Brazil with the exiled Portuguese King João VI in 1808 (Pardal,1985). The second School to be established was the Ouro Preto School of Mines (in 1876) which instituted a course on mining and metallurgical engineering.

Other schools soon followed: the Pernambuco Engineering School (1895), the Mackenzie Engineering Schools in São Paulo (1896), the Porto Alegre Engineering School (1896), the *Escola Politécnica* of Bahia (1897), the Belo Horizonte Free School of Engineering (1911), the Paraná Engineering School (1912), the *Politécnica* of Recife (1912), the Itajubá Electrical Engineering and Technical School (1913), the Juiz de Fora Engineering School (1914), the Military Engineering School in Rio de Janeiro (1928) and, finally, the Pará Engineering School in 1931 (Telles, 1993).

The above schools aimed to train civil engineers to work in the burgeoning cities, where they would be responsible for topographical surveys, all types and sizes of public and private buildings, road systems, canals, water and sewage networks, as well as for the conservation, planning and budgetary details involved in the public works that were an inevitable product of the growth of Brazil's urban areas.

2.2 Consolidation

In February 1911 Eng. Victor da Silva Freire gave a keynote address at the Guild of *Escola Politécnica* of São Paulo in which he advanced a theoretical justification for the proposal which formed part of a series of *avant-garde* town planning projects submitted by the Municipal Works Management Division. This proposal focused on the need to respect fundamental artistic and traditional principles and the non-static nature of cities which, he believed, could be transformed by designing and applying specific street patterns (Freire, 1911). Freire, as Professor of Engineering at the *Escola Politécnica* of São Paulo, was a devotee of the International Congresses for City Construction, which he attended regularly in Europe.

According to Simões Jr. (2004), Freire was the first to introduce the concept of town planning to Brazil. He was also the first engineer to treat this as a science rather than as a straightforward technical approach to street planning (as had hitherto been the case). Freire was the first to introduce a heightened theoretical approach to the subject - an approach which was becoming increasingly employed in other parts of the world.

The principal influences at the time were three European urban experts: Camillo Sitte (1843-1903, Austrian), Joseph Stübben (1845-1936, German) and Eugène Hénard (1849-1923, French). All these were considered to be the forerunners of modern 'urban science'. In addition to these three, the influence of the Englishman Raymond Unwin (1863-1940), was also notable. Unwin was responsible for Cia City in São Paulo (1912) built on the lines of the Garden Cities concept formulated by Ebenezer Howard. Ebenezer Howard (1850-1928) put forward the idea of building new cities with factories and gardens, The Garden Cities with houses built near to workplaces and the city center and within easy reach of green space. One of the main features of this design concept was the layout of the road and street systems which generally followed existing topography, however hilly or winding, thereby creating a more 'natural' environment.

Sitte, author of "Der Städtebau nach seinen künstlerischen Grundsätzen" (Building cities based on artistic principles) was a harsh critic of Haussmaniana (the 'grand monumentalist' approach), preferring to think in terms of irregular and more artistically- inspired patterns of streets and public squares. Baron Haussmann (1809-1891) was responsible for the rehabilitation of parts of the city of Paris by planning major thoroughfares, laying down fine parks and erecting a number of prestigious public buildings. Stübben, author of "Der Städtebau" (The building of cities) was, on the other hand, primarily concerned with questions of urban growth and issues touching on radial (spoke) and circumferential arterial road systems, as well as building healthy environments and promoting keener awareness of aesthetic factors. Hénard, author of "Études sur les transformations de Paris" (Studies on transforming Paris), produced a number of solutions for developing and improving cities in the course of his comparative work on the urban development of Paris, Moscow, London and Berlin.

The word "urbanism" was employed for the first time in Brazil by Freire (1916). This is a neologism of the French term *urbanisme* which emerged earlier in the century (in 1910) and which in turn was a translation of the English term 'town planning' (used for the first time in England in 1906). Similar terms had already been employed in Germany since the mid-19th century: *stadtplan* (city plans) and *stadtbau* (city building). Thus 'urbanism', or town planning, evolved into a modern urban science, reflecting the need to introduce a degree of planning discipline as the result of the major changes taking place in cities caused by industrialization and rapid population growth (Choay, 1965).

According to Freitag (2006), only with the advent of Le Corbusier (1887-1965) considered to be the founding father of modern town planning, could "urbanism" be considered to have become a universally accepted science, capable of providing practical solutions to the urban problems emerging in the context of 20th century industrial society.

The first 'urbanists' in São Paulo were civil and architectural engineers. These individuals left a clearly identifiable mark on the first examples of urban engineering in the growing city despite opposition from local administrators schooled not in engineering but in the law such as João Theodoro, Antonio Prado and others.

This group of urban engineers was educated at the *Escola Politécnica* (where a number of them also taught). They tended to align themselves with Victor Freire and his assistant - Eng. João Florence Ulhôa - who in 1924 conceived the idea of the "radial roads perimeter" and who in 1930 published, together with Eng. Francisco Prestes Maia, the first major street plan (the *Plano de Avenidas*) for the city of São Paulo. Eng. Prestes Maia, professor at the *Escola Politécnica,* and Mayor of São Paulo on two occasions (27 April 1938 - 27 October 1945 and 10 April 1961-7 April 1965), was considered by Toledo (1996) to be a major proponent of town planning strategy and doctrine, with a reputation as a tough administrator.

It is also worth mentioning the important roles played by Arthur Saboya and Francisco Rodrigues Saturnino de Brito (the latter known primarily for his work as a public health specialist) and Luis Ignácio Romeiro de Anhaia Mello, who belonged to the new generation of engineers greatly influenced by the new approach to urbanism in the United States.

Anhaia Mello was the main force behind the creation of São Paulo´s Architecture and Town Planning Faculty in 1948 - an independent academic facility which emerged from the engineering and architecture course previously run by the *Escola Politécnica.* Mello was the first director of this faculty and was primarily responsible for perceiving the inter-related aspects of "urbanism" and "architecture" (hence the name of the new faculty). (Ficher, 2005).

At the time the above engineers were working in São Paulo (the first half of the 20th century), the city underwent a major period of expansion which, in turn, justified the increasing concern directed towards town planning matters. Table 1 contains population data for 1872-1950.

Year	São Paulo Municipality			Brazil	
	Population	Annual geometric growth rate	Urbanization rate (%)	Population	Annual geometric growth rate
1872	31,385		-	10,112,061	
		4.1			2.0
1890	64,934		-	14,333,915	
		14.0			1.9
1900	239,820		-	17,318,556	
		4.5			2.9
1920	579,033		-	30,635,605	
		4.2			1.5
1940	1,326,261		94.9	41,236,315	
		5.2			2.3
1950	2,198,096		93.4	51,944,397	

Table 1. Population figures (IBGE, Demographic Census)

Souza (2006) notes that throughout this period large numbers of São Paulo Polytechnic engineers occupied public positions in the various municipalities and public works/road and street planning secretariats, with the majority of them closely involved in urban engineering activities.

The aforementioned urban engineers tended to regard the city as a whole unit - an approach which in their view called for integrated interventions of a technical and aesthetic nature with regard both to buildings and traffic organization. They also paid strict attention to the public sanitation aspects of the city in their plans for city streets and squares. Furthermore, they took into account the administrative and management aspects of the city, resulting in the establishment of a number of bodies employing specialist professional staff concentrated specifically on town planning.

The above professionals were mainly 'civil' or 'architectural' engineers who on graduating were attracted by the prospect of interesting, well-paid and prestigious jobs in this area of expertise.

The term 'urban engineering' was employed by Francisco de Paula Dias de Andrade in his thesis dated 1966 (Chair (*Cátedra*) No. 12: Buildings construction; Notions of architecture; Urban engineering and urbanism), submitted as part of the qualification process for a senior professorship appointment at the *Escola Politécnica*. Regardless of the fact that subsequent documents written by Professor Andrade fail to cast more precise light on the prospects for urban engineering in São Paulo, it is nevertheless evident that Andrade showed a keen pioneering approach with his creation in 1970 of a graduate course in construction and urban engineering at the *Escola Politécnica* of University of São Paulo devoted specifically to training engineers at Masters and Doctoral level in those fields of knowledge.

3. Urban engineering

According to Martinard (1986), urban engineering can be described as "the art of conceiving, undertaking, managing and coordinating the technical aspects of urban systems. The term 'urban technical systems' has two meanings: the first conveys the 'physical' dimension of an infrastructural 'support' network, while the second can be construed as a supporting 'services' network". For example, while the water supply system of any city possesses a 'physical' dimension insofar as the actual physical distribution of water is concerned (pipes, water capture machinery, treatment equipment etc), it is vital to take into consideration, in addition, the number and quality of the services required to operate and maintain the networks and their various equipments, to ensure appropriate billing, charging and cost recovery mechanisms for the payment of services rendered and the need for water quality control and supervision of the multifarious aspects of systems management.

It could be argued that the responsibility for the purely technical aspects of water supply falls to civil engineers specializing in hydraulic and sanitation engineering - a speciality widely recognized as one of the most traditional branches of engineering. However, although this particular class of engineer is certainly qualified to deal with and resolve problems in his chosen area of expertise (hydraulics and sanitation) it is difficult to attribute to him the title of 'urban engineer'.

A further example is that of the civil engineer specializing in transport engineering. This branch of engineering involves dealing with land, maritime, river and air transport, as well as the infrastructure needed to keep abreast of developments in these specialist areas. It is equally difficult to describe this transport specialist an 'urban engineer'.

Both the above examples point to the need to identify a more precise definition of the 'urban technical systems' mentioned by Martinard, given that there is no clear distinction made in current day-to-day practice between specialist civil engineering fields and those specifically associated with 'urban engineering'.

A further definition of the term is provided by EIVP, the École des Ingénieurs de la Ville de Paris (City of Paris Engineering School, http://www.eivp-paris.fr/), founded in 1959, which runs an undergraduate course in urban engineering. For the EIPV urban engineering deals with the 'conception, construction and management of cities', while simultaneously playing close attention to the need for 'sustainable development'.

In Anglo-Saxon countries, particularly in the United Kingdom, Canada and the United States, the term "municipal engineering" has a similar meaning to "urban engineering". Municipal engineering includes all the civil and environmental engineering services related to the complex problems generated by infrastructural and environmental problems and land use that confront municipal governments on a daily basis (see http://www.nlja.com/municipal.html). In our view, this more precise definition gives a clearer idea of the practical scope of urban engineering and of the activities undertaken by urban engineers.

Based on this definition, urban engineering can more properly be described as the branch of engineering that covers all the civil and environmental engineering services related to the range of complex problems associated with infrastructure, services, buildings, environmental and land-use issues generally encountered in urban areas.

3.1 The systemic approach

The 'urban engineer' operates in a broad and systemic manner, given that his field of activity is multifaceted and complex, involving many different social, economic, political, environmental and technological factors. This generally means that a large number of interests and stakeholders are involved.

Systemic (or systems) thinking is a framework based on the belief that the component parts or properties of an organism or living system can best be understood in the context of relationships with each other and with other systems rather than in isolation. These properties are the product of a variety of interactions and relationships between the separate parts and it follows that the only way to fully understand why a problem occurs and persists is to understand the part in relation to the whole (Ackoff, 1974).

This approach is of crucial importance if we wish to understand our cities and find ways of tackling the problems incurred in and by these cities. The many problems, for example, encountered in cities linked to water and energy supply, transport, etc cannot be seen in isolation. Rather they need to be understood systemically within the context of an overarching, broader urban context.

Applying this approach is a complex task given that urban engineering touches on a wide range of activities, including:

- water resources engineering, the collection and treatment of sewage, solid waste management, collection and disposal, energy distribution, drainage, urban transport, telecommunications, etc;

- different areas of activity requiring coordination, including initial planning and detailed design, project execution, ongoing operation, maintenance and management;
- a variety of stakeholders: public authorities at the several levels (local, municipal, regional, metropolitan, state and national), plus para-statal sectors, private sector involvement, NGOs and community representatives.

3.2 The urban space

One key aspect worthy of consideration is the relationship between the urban engineer and his territorial preserve - which amounts in reality to the entire "urban space". This particular space is not merely an area of land distinguishable from the "rural space" around the city but also a political space which also provides a home for members of the workforce (Castells, 1983).

However we believe that the urban and rural territorial spaces are in fact closely integrated and cannot be viewed as totally independent entities. In an urban engineering context it is thus vital to employ the 'systemic' approach in an effort to understand the mutually dependent relationship between the city and the surrounding country areas (and vice versa).

We also need to remember that any definition of what exactly is "urban space" tends to be fairly arbitrary. In Brazil, areas defined in municipal law (based upon the National Taxation Code) are considered to be 'urban areas'. The National Taxation Code defines those areas which can be considered as 'urban'. This definition is directly related to the IPTU (urban property rates) revenue. It is not necessary for these areas to be occupied by a minimum number of inhabitants or to possess a minimum population density. Other criteria have of course been adopted in other countries according to their homegrown political, geographic or cultural circumstances (Jenkins et al., 2007)

In the United States, for example, an area is considered to be "urban" when it has a minimum of 2,500 inhabitants, with a minimum population density of 1,000 persons other per square mile (386 persons per km^2 (one square mile = approx. $2.59km^2$)). This figure is similar to that applying to Canada, where a minimum of 1,000 persons per 4 km^2 is the norm. In Mexico an urban space requires at least 2,500 inhabitants but no density requirement. On the other hand, in Peru any area demarcated formally as "urban" has to possess *in situ* at least 100 dwelling units.

According to Veiga (2002) if the criteria of population size were combined with local and demographic density, the urban part of Brazil would represent 57% of the total population of the country (in 2,000) - not 81.2%, as stated by the Brazilian Statistics Institute IBGE.

4. The challenges

Practitioners of urban engineering are currently faced by the highly complex situation outlined above. Cities of different sizes and social/political weight are crying out for specialized and competent engineers possessing broad managerial expertise combined with a systemic approach to the tasks in hand. Cities are complicated environments requiring the involvement of fully qualified professional staff capable of confronting the many challenges, particularly in the cities of the developing world.

University-level urban engineering teaching in Brazil has traditionally been carried out at graduate level. The following urban engineering graduate courses were registered according to their original date of introduction: 1970, USP *Escola Politécnica*; 1994, Federal University of São Carlos; 2000, Federal University of Paraíba; 2002, Federal University of Uberlândia and Federal University of Passo Fundo; 2005, Federal University of Bahia; 2006, Federal University of Maringá; 2008, Federal University of Rio de Janeiro and Rio de Janeiro Catholic University. Two undergraduate courses have also come to our notice: one at the Federal University of São Carlos and the other at the ABC Federal University.
We need to train and qualify our urban engineers to face, inter-ally, the following challenges:

(a) ever-growing urban population pressure on existing infrastructure and public services. As can be seen in Table 2, "macro-regions" throughout the world have recorded continuing urban demographic growth in both absolute and percentage terms;

	Population (millions)			Urban population (millions)			Urban population (%)		
	1950	2005	2030	1950	2005	2030	1950	2005	2030
World	2,535	6,464	8,200	735	3,148	4,912	29.0	48.7	59.9
Africa	224	922	1,518	33	353	770	14.7	38.3	50.7
Asia	1,410	3,938	4,931	237	1,567	2,668	16.8	39.8	54.1
Europe	548	731	707	277	528	554	50.5	72.2	78.3
LA & Caribbean	168	558	713	71	432	601	42.0	77.4	84.3
North America	171	332	405	109	268	351	63.9	80.7	86.7
Oceania	13	33	43	8	23	32	62.0	70.8	73.8

Table 2. Demographic change by macro-region (UN World Population Prospects, 2006)

(b) in spite of this growth, what have been observed is that in general the quality of living in cities improves as increases the urbanization rate, particularly in developing countries. As an example of this phenomena, Figure 1 shows that as the urbanization process advances, the infant mortality rate which is a largely adopted social indicator, decreases. It is quite logical that this kind of situation occurs because population will have more access to health care, education and information in cities even if these services are not so well delivered. This situation leads to an approach which understands cities not only as a problem but the solution, or at least an important part of it.

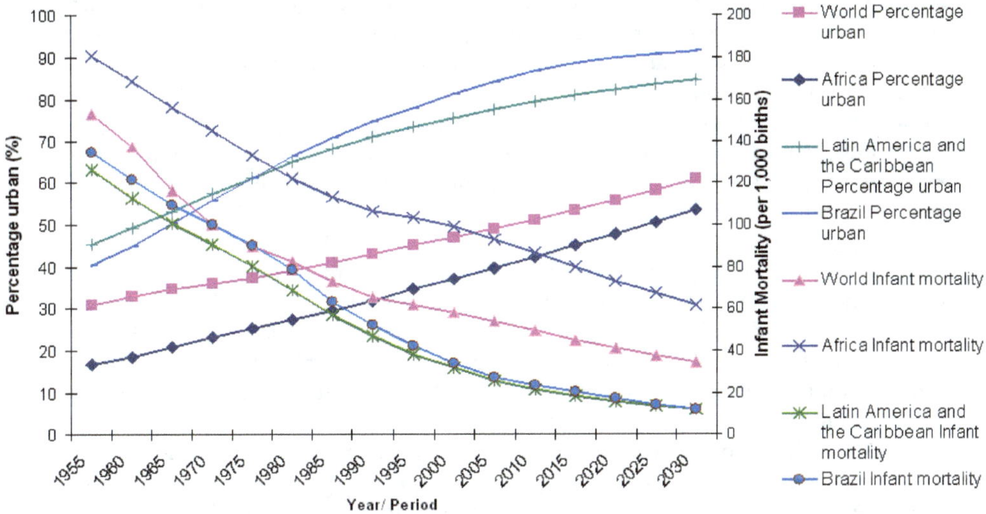

Percentage urban. Source: United Nations Population Division (2006). World Urbanization Prospects (2005). Infant mortality. Source: Spreadsheets provided by the United Nations Population Division. Both genders combined.

Figure 1. Percentage urban versus Infant mortality in World, Africa, Latin America and Caribbean, and Brazil

(c) urban population growth has tended to be concentrated on the metropolitan regions, given that these attract incoming workers to available employment opportunities. Brazil's 'metropolitan regions' possess formal and legal status but they are not in reality 'political' entities as such, able to exclusively benefit from government resources while undertaking appropriate responsibilities and commitments. This creates serious administrative difficulties since problem-solving is often not confined to the territorial boundaries of a particular municipality but calls for intervention at a wider regional level. A prime example of this situation is the whole question of the final disposal of solid waste;

(d) the deterioration and obsolescence of existing infrastructure networks and the need to introduce new technical solutions in keeping with the physical and population growth of the cities. In this respect many new, lighter and more durable materials have come onto the market but these have often not been properly tested in real situations. Moreover, the higher building densities in the urban areas ("verticalization") generally mean that infrastructure and services networks need upscaling in order to meet new demands;

(e) the introduction of new technologies such as cellphones and the internet and the rapid evolution of increasingly more efficient, accessible and cost-effective information management, access and retrieval systems such as those based on geo-processing;

(f) complex, decentralized and automated administrative and governmental systems requiring efficient and coherent coordination and follow-up. Financial resources are under massive pressure everywhere, calling for the development of efficient 'allocation and usage' criteria by urban management practitioners. The need for maintaining good lines of communication with members of society and organized economic sectors is also important. While it is obvious that the interests of these urban stakeholders have to be taken into proper consideration, the broader interests of society as a whole need to be respected in the short and, above all, the longer term, with due attention paid to the relevant strategic planning processes;

(g) increased community participation demanding a higher level of transparency on the part of the public authorities. Communities have begun to protect their own interests at the neighborhood and city block level by employing direct action, as well as through indirect pressure exerted by social organizations. Communities have also expanded the scope of their activities and are currently in a better position to influence, for example, master plans and other urban planning laws at the initial stages. It is also worth mentioning that professional and corporate associations are increasingly involved in pressuring local authorities to undertake appropriate action. The latter, for their part, are increasingly obliged to engage their interlocutors in sensible dialogue;

(h) increasing involvement by the private sector through concessions and permits which call for complex bidding, tendering, contracting, control and remuneration systems. The so-called Public Private Partnerships (PPPs) currently provide new opportunities for service provision and the sharing of responsibility between public-sector and private bodies;

(i) the growing need for the processes and products developed and used in cities to comply with environmental requirements. These requirements, apart from conforming to new compulsory environmental legislation, are also the outcome of a series of social demands presented by NGOs, community groups and by the many proactive voluntary approaches by private service delivery organizations. Also on the environmental level, it is worth noting the increasing inroads made by systems that govern the rational use of water and energy contributing to reducing global warming. In this aspect it is important to register the importance of the urban transportation as one of the main responsible for the environmental problems which affect contemporary cities.

A further crucial challenge exists in many developing countries: problems arising from the contiguity of conventional, "formal" cities with "clandestine", "informal" cities. Given their size, the latter - consisting mainly of *favelas* (slums) and irregular subdivisions - can no longer be considered as illegal settlements, mainly on account of their large size. According to Benevolo (2006), past attempts to suppress the informal areas of cities (replacing them with planned developments and/or relocating the inhabitants) have met with limited success. It is now generally accepted that in the longer term the best way to approach this situation is to introduce incremental improvements and to stabilize the original irregular land occupations by introducing basic infrastructure and services to the poorer areas in question.

In Brazil this approach is perhaps best illustrated by the slum upgrading (*urbanização de favelas*) initiatives that are being taken forward in the majority of our large cities. Moves are afoot to retain the resident populations in the already-occupied areas while improving living conditions by introducing better street layouts, eliminating risk areas, installing water supply and sewage/storm-water collection systems and electricity/telephone distribution networks, street-lighting etc. A range of other public services and complementary facilities such as income generation and post-works social monitoring programs have frequently gone hand in hand with public works in these problematic areas (Abiko, 2007). Some of the *favelas* have in fact become 'real' cities, in view of their enormous size and number of inhabitants (Marques (2007) has produced an interesting survey of "precarious settlements" in Brazil).

The *favela* upgrading developments have involved the participation of architects, lawyers, social workers, doctors and engineers, together with other professionals working in interdisciplinary teams. It is now obvious that in housing interventions of this nature the involvement of the *urban engineer*, possessing a clear understanding of systemic urban requirements and an ability to act accordingly, is paramount. The services of the urban engineer are vital not only at the project design, planning and execution level but also at the technical and 'social' levels - two specific areas of expertise that go beyond the traditional narrow confines of the qualified civil engineer's job description.

To conclude, it is clear that engineers with a broad, systemic approach rooted in the historic efforts of the pioneering urban engineers at the beginning of the last century, have an extremely important future role to play in our cities. Although the urban engineering pioneers labored in totally different circumstances a century ago they nevertheless continue to serve as examples of clear-sightedness and dedication in the quest for a better quality of life for the inhabitants of our cities.

5. References

Abiko, A. et al. (2007). Basic Costs of Slums Upgrading in Brazil. *Global Urban Development Magazine*, Vol. 3, Issue 1, Nov 2007, Washington. (http://www.globalurban.org/GUDMag07Vol3Iss1/Abiko.htm)

Ackoff, R. L. (1974). *Redesigning the future: a systems approach to societal problems*, Wiley, New York

Andrade, F. P. D. (1966). *Subsídios para o estudo da influência da legislação na ordenação e na arquitetura das cidades brasileiras*. Escola Politécnica da USP, Tese de Cátedra, São Paulo

Benevolo, L. (2006). *L´architettura nel nuovo millenio*. Laterza & Figli SPA, Roma/Bari

Castells, M. (1983). *A questão urbana*, Paz e Terra, Rio de Janeiro

Choay, F. (1965). *L' urbanisme: utopie et réalités*, Seuil, Paris

Ficher, S. (2005). *Os arquitetos da Poli: ensino e profissão em São Paulo*, Edusp, São Paulo

Freire, V. S. (1911). Os Melhoramentos de São Paulo. *Revista Polytechnica*, (33): 100, fev./mar. 1911, São Paulo

Freire, V. S. (1916). A planta de Bello Horizonte. *Revista Polytechnica*, 9 (52): 159-174, São Paulo

Freitag, B. (2006). *Teorias da cidade*, Papirus, Campinas

IBGE, Instituto Brasileiro de Geografia e Estatística. (http://www.ibge.gov.br)

Jenkins, P., Smith, H., & Wang, Y.P. (2007). *Planning and housing in the rapidly urbanizing world,* Routledge, London

Marques, E. (Org.) (2007). *Assentamentos precários no Brasil urbano.* Ministério das Cidades e Centro de Estudos da Metrópole. Brasília/São Paulo

Martinard, C. (1986). *Le génie urbaine,* Rapport au Ministre de l´Équipment, du Logement, de l´Aménagement du Territoire et des Transports. La documentation française, Paris

Pardal, P. (1985). *Brasil, 1792: Início do ensino da engenharia civil e da Escola de Engenharia da UFRJ,* Fundação Emílio Odebrecht, Rio de Janeiro

Santos, M. C. L. (1985). *Escola Politécnica da Universidade de São Paulo: 1894-1984,* Reitoria da USP/FDTE, São Paulo

São Paulo (cidade). Prefeitura do Município. (1950). *Ementário de legislação referente ao período 1890-1950.* São Paulo

Simões Jr., J. G. (2004). *Anhangabaú: história e urbanismo.* Editora Senac e Imprensa Oficial do Estado de São Paulo, São Paulo

Souza, A. C. R. (2006). *Escola Politécnica e suas múltiplas relações com a cidade de São Paulo, 1893-1933,* PUC/SP, Doutorado, São Paulo

Telles, P. C. S. (1993). Duzentos anos de ensino de engenharia no Brasil. *Revista Politécnica,* No. 211, ano 90, outubro/dezembro 1993, São Paulo

Toledo, B. L. (1983). *São Paulo: três cidades em um século,* Duas Cidades, São Paulo

Toledo, B. L. (1996). *Prestes Maia e as origens do urbanismo moderno em São Paulo,* Empresa das Artes, São Paulo

United Nations World Population Prospects, The 2006 Revision at http://esa.un.org/unpp/

Veiga, J. E. (2002). *Cidades imaginárias: o Brasil é menos urbano do que se calcula,* Editora Autores Associados, Campinas

6

Urban Noise Pollution Assessment Techniques

Fernando A. N. Castro Pinto
Federal University of Rio de Janeiro (UFRJ)
fcpinto@ufrj.br
Brazil

1. Introduction

An important factor for the life quality in urban centres is related to the noise levels to which the population is submitted. Several factors interfere with the amount of noise pollution throughout the city. Among them, and as one of the most important, is the traffic noise.
A major challenge is the quantification of the noise effects on the population. Not only high levels must be assessed but also the amount of people exposed to them is of great importance. This task is far from obvious since the sound propagation is affected by many environmental characteristics distinct in nature. The topology of the buildings and the topography may create quiet zones even in crowded neighbourhoods. Traffic may statistically vary. The population exposed might be resident but also fluctuate, not to mention the subjective nature of the sound perception itself.
In order to aid the urban planner cope with these difficulties, this chapter will comprehensively presents alternatives ranging from numerical simulation, called noise mapping, to measurement based noise monitoring.
Noise mapping techniques together with standards for the calculation of noise propagation are powerful tools to aid urban planners in correctly applying noise abatement measures in an economically feasible way. Nevertheless the results of such mappings rely on a great amount of data, location and strength of noise sources, ground geometry, location and geometry of buildings, etc. This work also discusses the sensitivity of the obtained simulated noise levels to the quality and precision of the geometric data available.
Actual measurements are however needed not only to verify the model assumed for the simulation but also for the noise pollution assessment itself. This can be achieved through local measurements of short duration or through long term monitoring in fixed places. The measurement techniques and procedures are addressed together with the creation of databases to help the decision making process of the urban planner

2. Sound Propagation and Topology

A noise map is a tool that delivers visual information of the acoustic behaviour of a geographic area either in a specified moment or in a statistical base. It is considered as tool to improve or to preserve the quality of the environment regarding noise pollution, allowing a comprehensive look at the problem of multiple sources and receivers.

Noise map is also an excellent tool for urban planning. According to Santos (2004), the use of noise maps techniques as a planning tool allows:

- Quantification of noise in the studied area;
- Evaluation of the population exposition;
- Creation of a database, for urban planning with localisation of noisy activities and mixed and sensible zones;
- Modelling of different scenarios of future evolution;
- Prediction of impact noise of projected infrastructure and industrial activities.

In Europe, the Directive 2002/49/EC of the European Parliament and of the Council, of 25 June 2002 relating to the assessment and management of environmental noise imposes to its Member States the elaboration of noise maps for cities with more than 250,000 inhabitants, due no later than 30 June 2007 (EC, 2002). These maps shall be reviewed, and revised if necessary, at least each five years after the date of their preparation. In Brazil, however, the presentation of noise maps by the city planners is still not an obligation. In Rio de Janeiro, specifically the local legislation, supported by the corresponding federal one, only foresees maximum acceptable levels of noise according to the occupation type or urban zone.

The elaboration of maps can be made using real measurements in points previously determined, using only prediction models through simulations or, in a mixed system, simulations can be complemented and verified with actual measurements. Of course the core of a noise map resides in the propagation model of the sound originating by the sound sources, and the model used for these sources itself.

The propagation model must take into consideration the usually high concentration of population, shops and a heavy traffic from particular vehicles and public transportation, in a general urban environment. Of course there are considerable differences between neighbourhoods of a big city, densely populated, and small city with lesser buildings and more free area. Although the result of the propagation of sound being quite different in these cases the mathematical model behind the calculations is the same. It must consider the effect of the ground topography, the presence of natural or artificial barriers, the effect of reflection and diffraction of the sound waves on buildings and facades but also on the ground itself. For the majority of commercially available software the propagation model is defined in national standards, which are incorporated in the calculation code. Table 1 lists some commonly found standards, from different countries, that establishes noise calculation procedures. Not only the propagation but the modelling of the sound generation is included, depending on the kind of source being simulated (Datakustik, 2005).

In this way, not only the results may be verified independently, but also the noise map can be presented according to the corresponding local legislation enforcing specific standards. Of course one still need to chose one of the available standards to perform the calculations for the case where no specific model is required (City of Rio de Janeiro, 1978, 1985, 2002, and ABNT, 2000).

The topography of the region is input to the software either as basic data from a CAD model or through the use of a aerial photographic image of the desired area with the corresponding terrain heights input manually. Usually, CAD database do not include only the topography of the neighbourhood under study, but also the individual building heights.

This kind of information may be available for the majority of great cities, otherwise the cost of a simulation will increase with the increase in time to input the data. Figure 1 shows the computer representation of the topography of the terrain, including also the buildings with their individual properties of an area under study (Pinto et al., 2005).

Type of Source	Standard or Calculation procedure
Industrial Noise	ISO 9613 incl. VBUI and meteorology according to CONCAWE (International, EC-Interim) VDI 2714, VDI 2720 (Germany) DIN 18005 (Germany) ÖAL Richtlinie Nr. 28 (Austria) BS 5228 (United Kingdom) General Prediction Method (Scandinavia) Ljud från vindkraftverk (Sweden) Harmonoise, P2P calculation model, preliminary version (International)
Road Noise	NMPB-Routes-96 (France, EC-Interim) RLS-90, VBUS (Germany) DIN 18005 (Germany) RVS 04.02.11 (Austria) STL 86 (Switzerland) SonRoad (Switzerland) CRTN (United Kingdom) TemaNord 1996:525 (Scandinavia) Czech Method (Czech Republic)
Railway Noise	RMR, SRM II (Netherlands, EC-Interim) Schall03, Schall Transrapid, VBUSch (Germany) Schall03 new, draft (Germany) DIN 18005 (Germany) ONR 305011 (Austria) Semibel (Switzerland) NMPB-Fer (France) CRN (United Kingdom) TemaNord 1996:524 (Scandinavia) FTA/FRA (USA)
Aircraft Noise	ECAC Doc. 29, 2nd edition 1997 (International, EC-Interim) DIN 45684 (Germany) AzB (Germany) AzB-MIL (Germany) LAI-Landeplatzleitlinie (Germany) AzB 2007, draft (Germany)

Table 1. Parameters needed for a noise impact study through a map

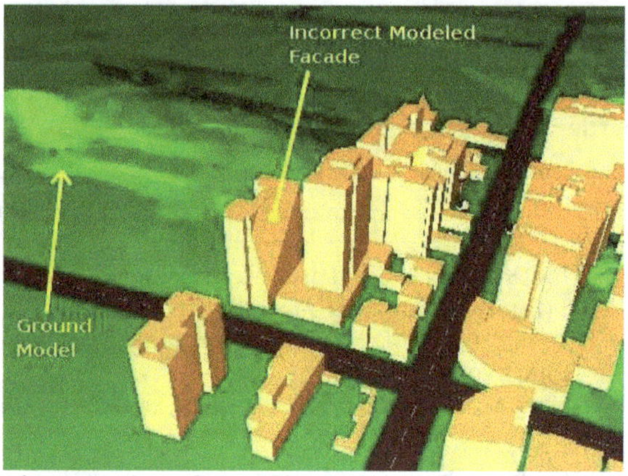

Fig. 1. Topography of a region under study with terrain and building elevations (only a partial number of buildings is depicted)

As a next step after the topological information is correctly inserted into the software database, which can be done in a very automated way from CAD programs, the noise sources must be identified and modelled. Several commercial software can be used to calculate noise maps, among them may be cited CADNA-A, Mithra, SoundPlan, Predictor, IMMI, LIMA, ENM, etc. To create the noise maps presented in this work the software CADNA-A was used. The modelling, following the procedures established in the standard being used, is based on different parameters (Table 2).

<table>
<tr><td rowspan="6">Source</td><td rowspan="3">Traffic noise</td><td>Type of vehicles (car, motorcycle, truck)</td></tr>
<tr><td>Type of engines (gasoline, diesel)</td></tr>
<tr><td>Mean velocity</td></tr>
<tr><td colspan="2">Industrial noise</td></tr>
<tr><td colspan="2">Rail noise</td></tr>
<tr><td colspan="2">Entertainment</td></tr>
<tr><td rowspan="4">Surroundings</td><td colspan="2">Road surface</td></tr>
<tr><td colspan="2">Building heights</td></tr>
<tr><td colspan="2">Street widths</td></tr>
<tr><td colspan="2">Absorption coefficients (facades)</td></tr>
<tr><td rowspan="3">Environment</td><td colspan="2">Humidity</td></tr>
<tr><td colspan="2">Temperature</td></tr>
<tr><td colspan="2">Wind</td></tr>
<tr><td rowspan="2">Demographic parameters</td><td colspan="2">Number of inhabitants</td></tr>
<tr><td colspan="2">Number of units per building</td></tr>
</table>

Table 2. Parameters needed in a noise impact study

For instance when dealing with traffic noise the propagation is characterised by diverse parameters (type of vehicles, number of vehicles) and surroundings (height of the building, sound absorption coefficient of the facade, type of floor, width of the streets) influencing in noise propagation. Actually we can distinguish between a small number of source types (Kinsler et al., 1982):

- point source (like a loudspeaker, a valve, a vehicle, an aeroplane, an operating industrial equipment, etc.);
- line sources (like a road, a railway, piping system, etc.);
- area sources (like a parking lot, people gathering together, the openings of a tunnel, etc.);

which will be most basically modelled by their sound power. Table 3 shows the source of information for parameters.

Parameter	*Source of Information*
Terrain topography	Maps, CAD-models, Aerophotos, Satellite Images
Position and dimensions of buildings	Maps, CAD-models, Aerophotos, Satellite Images
Height of buildings	CAD-models, Field Information
Type of facade absorption	Field Information
Position and dimensions of noise barriers	CAD-models, Field Information
Height of barriers	CAD-models, Field Information
Position and cross section of roads	CAD-models, Field Information, Traffic Management
Traffic volume in roads	On-Line Information Systems, Traffic Management, Video Systems, Manual or Automated Counting
Percentage of heavy vehicles	Traffic Management, Video Systems, Manual or Automated Counting
Average vehicle speed	On-Line Information Systems, Traffic Management
Type of road paving	Traffic Management, Field Information
Sound power of generic sound sources	Direct Measurements, Equipment Specifications, Noise levels
Position of generic sound sources	CAD-models, Field information, Aerophotos, Satellite Images
Directivity	Direct Measurements, Equipment Specifications
Population density	Field Information, County Databases

Table 3. Source of information for parameters

The sound pressure levels produced by a sound source can not be considered an intrinsic characteristics of the source itself. The levels are rather a consequence of the interaction of the acoustic energy being introduced into the environment and the environment itself. It can be easily understood if one considers a loudspeaker operated in a well absorptive room like a studio compared with the same loudspeaker, fed with the same power, in a highly reflective environment like a bathroom. In the latter the reflection of the energy in the walls contribute to the sound level inside the room, whereas in the former the walls retains most of the energy, thus causing a smaller level.

Sound power, although in some circumstances being also influenced by the environment, can be regarded as a characteristics of the source itself and can be measured with different, standardised, procedures (ISO, 1994).

Starting from these data the program calculates the noise map of the selected zone. Nevertheless many factors may affect the correctness of the results obtained, i.e. of the model used. In order to validate the calculation, the simulated values from sound pressure levels should be compared with experimental measurements.

Since it can be expected that the noise predictions based on the German regulation RLS-90 would not match, for instance, the Brazilian vehicle fleet conditions this comparison is a primary issue. Based on the level differences between actual measurements and the simulation model, its parameters can be modified in order to get a better approximation of the real results by the simulation.

Firstly a general simulation of the neighbourhood noise levels is done, considering the volume of daily traffic, the average speed, the width of the streets, the type of asphalt, the sound power and location of other sources and the height of the buildings. To compare the values simulated with real measurements, a smaller sector may be considered in order to speed up calculations. With the simulation of the sector, the software generates a map of noise as shown in Fig. 2, which corresponds to the noise levels at a height of 1.5 meters, approximately the height of the measuring microphone. Table 4 shows a comparison between the simulation results and the real measured data (Pinto & Mardones, 2008).

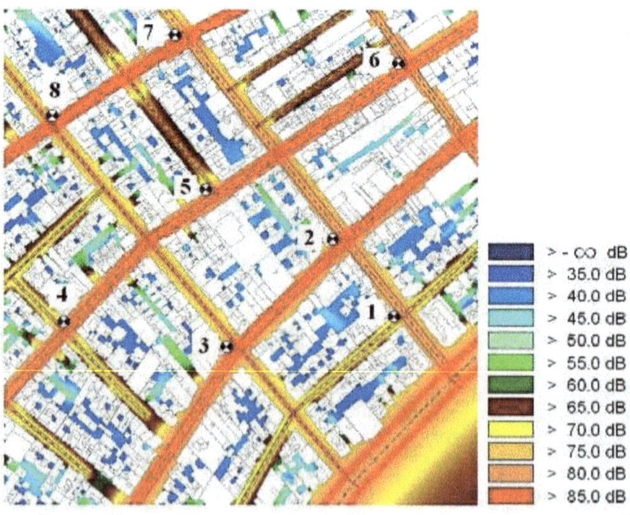

Fig. 2. Noise map of a small sector to compare with actual measurements (only traffic noise)

Point	Position	Measurement dB(A)	Simulation dB(A)	Difference dB(A)
1	Domingos Ferreira 76	65,1	65,7	-0,6
2	Domingos Ferreira/Figueiredo Magalhães	67,4	69,8	-2,4
3	Av. N.S.Copacabana 610	76	78,2	-2,2
4	Av. N.S.Copacabana/Figueiredo Magalhães	74,3	74,7	-0,4
5	Av. N.S.Copacabana/Santa Clara	73,5	73,6	-0,1
6	Santa Clara frente ao 98	70,5	70,6	-0,1
7	Av. Barata Ribeiro/Raimundo Corrêa	73,8	72,5	1,3
8	Av. Barata Ribeiro 535	74,8	76,7	-1,9
9	Av. Barata Ribeiro/Anita Garibaldi	71,8	73,3	-1,5
10	Av. Barata Ribeiro 432	77,6	77,4	0,2
11	Av. Barata Ribeiro/Siqueira Campos	73,6	75,6	-2
12	Rua Tonelero/Figueiredo Magalhães	71,7	75,8	-4,1
13	Rua Tonelero/Santa Clara	71,5	72,3	-0,8
14	Santa Clara 161	68,3	67,3	1

Table 4. Comparison between measurements and simulation after model correction

The parameters used in the simulation can then be modified in order to reduce the level differences obtained. It can be seen that the level difference is not the same at all positions, thus it may be quite challenging to try to adapt the model to meet all results in every situation. A lasting error of about 2dB or 3dB between measurements and simulation is therefore quite acceptable. Specifically for the case shown, which deals only with traffic noise, the vehicle volume at each street may be corrected to approximate the levels. This modification does not reflect bad information on the amount of traffic but rather the difference between the German and Brazilian vehicle fleets. Therefore it is advisable to verify the simulation, at least, in a restricted set of points, in order to adapt the sound source description to approximately reflect the measurements at these locations. After that more confidence can be inferred from the noise map obtained.

3. Mapping Results

The technique of noise mapping is a very powerful tool in urban planning. Not only the actual situation can be deeply studied but also, and probably the most important aspect, the noise pollution impact of every intervention of the city planners can be previously assessed. From a new layout of roads and avenues to the installation of an industrial facility, from new traffic orientation to the construction of a shopping mall, the sound pressure levels to which the population will be exposed can be determined from the model of the sound sources that may be considered. The necessary counter measures can be proposed and investigated in order to determine their effectiveness.

Although these studies are more commonly carried out in the process of identifying the environmental impact of major plants, like thermoelectrical power plants, their use should be extended and enforced to assess even the noise involved in the construction phase of an enterprise in a densely populated urban centre. Entertainment activities for a large number of people, ranging from shows in open spaces, like beaches, to the operation of a music club should be analysed in this way prior to official city approval.

Figure 3 shows a densely populated neighbourhood from the city of Rio de Janeiro, called *Tijuca*.

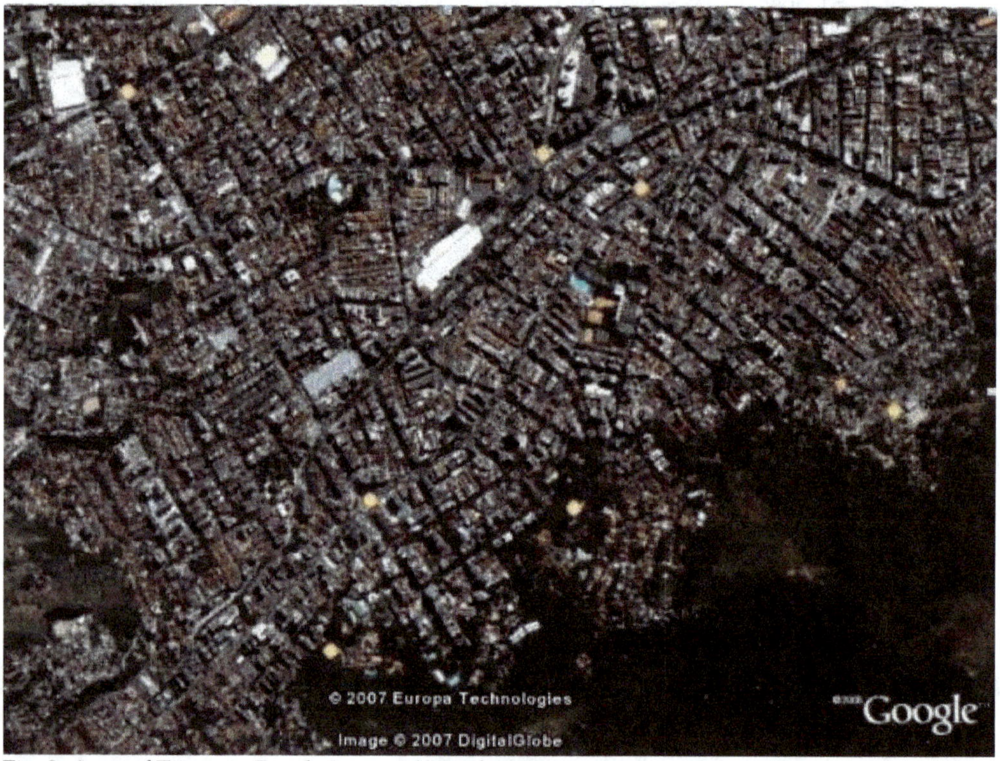

Fig. 3. Area of Tijuca in Rio de Janeiro (Google Maps)

A noise map study conducted in this area can be seen in Fig. 4, where the only source involved is the traffic noise. There are no remarkable sound sources of other kind in this area for it is a major residential neighbourhood.

It can be seen that the noise levels in the main avenues exceed tolerable limits, already due to the traffic noise alone. A reduction of municipal taxes for the most affected residences could be a first measure, if proposed in the city law, in order to bring the problem of noise pollution to attention of the administration.

Fig. 4. Noise map of Tijuca (only traffic noise)

Some open problems, specially in cities with an economical environment like Rio de Janeiro, is the quantification of the noise pollution in poor areas like the *favelas*. Coupled with that is the assessment of the noise impact from barely legal activities like popular music shows and parties (*Bailes Funk*) which are held in the favelas but affect the population both in the favela itself as well as in the regular city in the neighbourhood.

4. Conclusion

The assessment of noise pollution can be made through measurements which, however, are restricted to a limited number of points. The simulation of the sound waves propagation enables the study of a whole region in respect to the expected sound pressure levels as a result from existent sound sources. Of course, in order to perform a meaningful simulation, the environmental properties as well as the characteristics of the sound sources must be modelled. The results obtained may be gathered and presented graphically in a so called noise map. Actual measurements are used to verify and adjust the simulation to the real situation.

Specially in the case of urban centres noise maps allow the correct interpretation of the influence of distinct sources, the assessment of the sound pressure levels to which the population is exposed and the study of counter measures. The impact of major changes in the urban environment, like an industrial facility or a new road and traffic layout, can also be evaluated prior to implementation, together with the effectiveness of eventually proposed mitigation concepts.

The use of noise maps in the city planning is already incorporated in the European legislation but the Latin American, in general, and Brazil, specifically, laws can still be improved in order to enforce the compilation of noise maps and establishing goals to reduce the overall levels and the impact in the population. The noise map of a densely populated neighbourhood in Rio de Janeiro was presented.

5. References

ABNT - Assosiação Brasileira de Normas Técnicas (2000). *NBR 10151/2000 Acústica - Avaliação do ruído em áreas habitadas, visando o conforto da comunidade - Procedimento,* Rio de Janeiro, Brazil

City of Rio de Janeiro (1978). *Decree #1,601 from June 21st (1978),* Diário Oficial do Município do Rio de Janeiro, Brazil

City of Rio de Janeiro (1985). *Decree #5,412 from October 24th (1985),* Diário Oficial do Município do Rio de Janeiro, Brazil

City of Rio de Janeiro (2002). *Resolution #198 from February 22nd 2002 of the environmental board of the city,* Diário Oficial do Município do Rio de Janeiro, Brazil

Datakustik GMBH (2005). *CADNA Manual V3.4,* Greifenberg, Germany

EC (2002). *Directive 2002/49/EC of the European parliament and of the council of 25 June 2002 relating to the assessment and management of environmental noise,* Official Journal of the European Communities, L 189, pp. 12-26

ISO (1994). *ISO 3744 Acoustics - Determination of Sound Power Levels of Noise sources Using sound Pressure - Engineering Method in an Essentially Free Field Over a Reflecting Plane,* International Standards Organisation, Genève

Kinsler, L.E.; Frey, A.R.; Coppens, A.B. & Sanders, J.V. (1982). *Fundamentals of Acoustics,* John Wiley & Sons, New York, United States of America

Pinto, F.A.N.C.; Slama, J. & Isnard, N. (2005). *Sensitivity of noise mapping results to the geometric input data,* In: Rio internoise 2005/the 2005 Congress and Exposition on Noise Control Engineering, work #1847, Rio de Janeiro, Brazil

Pinto, F.A.N.C. & Mardones, M.D.M. (2008). Noise mapping of densely populated neighborhoods - example of Copacabana, RJ, Brazil, *Environmental Monitoring and Assessment,* on-line, doi: 10.1007/s10661-008-0437-9, to be published

Santos, L.C. & Valado, F. (2004). *The municipal noise map as planning tool,* Acústica, Guimarães, Portugal, Paper ID: 162

7

Locating Sites for Locally Unwanted Land Uses: Successfully Coping with NIMBY Resistance

Stefan Siedentop
University of Stuttgart
stefan.siedentop@ireus.uni-stuttgart.de
Germany

1. Introduction

Since the late 1970s, the Not In My Backyard ("NIMBY") phenomenon has become a challenge for urban planners, policymakers, developers and regulators in developed countries. NIMBYism characterizes a negative social response to locally unwanted land uses ("LULU's") such as nuclear waste repositories, land fills, mining activities, power plants, hazardous waste disposals or transport infrastructure. In addition, also social facilities like affordable housing projects, detention centers or homeless shelters can be objects of massive public resistance. Even "good" facilities that enjoy high support in general expressed in opinion surveys (e.g. wind energy farms or urban infill development) are faced with massive opposition at the "grassroots level".

For facility managers, it is a frustrating experience that many efforts to site a new development have ended in failure, in serious delays or in cost overruns due to unexpected mitigation and compensation measures. In contrast, local opponents often point to an unfair distribution of benefits and costs of "dirty" facilities and criticize technical dominated and hierarchical oriented planning approaches. What are the key factors that need to be considered and what are key mistakes that need to be avoided in a successful siting process of risk-related facilities? The intensity of public opposition against an unwanted facility or land use seems to be a function of three factors, the nature of the proposed activity, the nature of the region or community selected as a site, and the way in which the siting process is organized. Obviously, for engineers, urban and environmental planners, the latter is of crucial importance. Numerous studies show that the way in which urban planners and engineers deal with NIMBY attitudes held by local residents highly influences the viability of resistance and the outcome of planning. To effectively cope with social opposition against unwanted infrastructure, procedural fairness and transparency in planning is demanded.

Against this background, this chapter intends to enhance the understanding of the social aspects of the NIMBY phenomenon as a crucial prerequisite to effective and successful siting approaches. Planning recommendations are based on two "schools of innovation":

- the first considers siting as a social process of negotiating benefits and costs of unwanted land uses within a democratic decision arena;

- the second refers to an understanding of siting as a rational, knowledge-based process of systematically selecting site alternatives in a transparent manner.

The chapter is organized in three main parts. The first part (section 2) briefly describes the nature and social drivers of NIMBY responses. The literature related to the complexity of NIMBYism will be summarized. The other two sections introduce procedural standards as well as siting methods and techniques to carefully address NIMBY concerns. Section 3 portrays planning techniques to promote consensus building based on the results of social sciences. Section 4 describes innovative planning methods of siting "dirty" land uses based on standards of multicriteria decision making.

2. The Social Background of Negative Responses to Unwanted Land Uses

2.1 NIMBY concerns

Siting conflicts are extremely diverse in terms of the facilities that are considered as "locally unwanted", the locations involved, the impacts that might arise and the articulated reasons for opposition. Opponents of proposed facilities fear health risks and a decline in quality of life due to noise, traffic or threats to scenic beauty. An important motive is an expected decline in property values due to unattractive land uses in the neighborhood. Fischel (2001, p. 144) supposes that the rate of homeowners amongst LULU opponents is above average. Therefore, from a solely economic perspective, opposition is viewed as a rational "risk-averse" strategy by individuals who perceive a negative balance between the benefits they will gain from hosting a facility nearby, and the cost they have to bear (Lober 1995). Actually, there is clear evidence of a strong influence of spatial variables to the extent of NIMBY opposition. People who live closest to a facility are more likely to respond negatively. Lober (1995, p. 500) assumes an inverse relationship between distance and opposition. Individuals perceive lower risks form a planned facility the greater the distance they live from it.

However, NIMBYism cannot be narrowed to resistance of those people who live in the immediate affected area. Public opposition is also driven by people with broader interests in relation to environmental, social or political concerns. Opposition groups may include national-level non-government organizations that provide organizational and financial resources, based on their concerns about environmental protection, social justice or ethical standards (Schively 2007, p. 257). Sometimes, these groups do not only criticize the location of a noxious facility, but rather the principal demand of a technology or land use.

Another important issue is the different ways NIMBY responses are being expressed. Opponents often argue that the facility is not needed or does not belong in the area. The latter challenges land uses, which are not flexible in their location. This applies to resource extraction activities or renewable energy use. Here, "it is the site that chooses the project, not the reverse" (Kahn 2000, p. 22). Next to the question of location, NIMBY groups often criticize the siting process and the planning and participation procedures as insufficient, arbitrary or unfair.

In many cases NIMBYism resulted in the abandonment of the project under consideration. This can lead to a lack of access to needed services, associated with an excessive demand for transportation to receive service (e.g. transport of noxious waste to facilities in other states). Critics charge that NIMBYism has the potential to produce serious "gridlock situations" with negative effects on economic prosperity and social welfare. Another consequence of "successful" NIMBY opposition is the use of service strategies that are less efficient from an environmental or economic point of view, or the use of land that is not as suitable for siting a facility. Some scholars even assume that NIMBYism is jointly responsible for "leapfrogging" types of urban sprawl because planers and facility managers prefer sites with larger distances to "socially sensitive" land uses (e.g. residential areas) where less opposition can be expected.

2.2 Social and psychological drivers

The economic and social roots of NIMBY disputes can be explained by the specific spatial distribution of costs and benefits of an unwanted development (Lober 1995; Davy 1997). Lober (1995, p. 500) points to the fact that the net costs, though small to society, are relevant for individuals who live nearby the facility, thus stimulating NIMBY responses. In contrast, the net benefits to each member of the society are small, resulting in a limited incentive to politically support the facility. In other words, the regional benefits of a development are exceeding the local costs in total. However, from the perspective of an affected resident, the costs at the local level significantly outpace the low per capita benefits at the higher regional level. Table 1 provides a simple example to demonstrate this siting-intrinsic dilemma (Davy 1997): the operation of a hazardous waste facility is expected to yield a total benefit of 100,000 units of utility to a region with 1,000,000 residents. As a consequence of the development, 1,000 residents near the site incur a loss of 10,000 units of utility. In total, there is a net benefit of 90,000 utility units - without any doubt, the project would be profitable. For the 1,000 residents that are directly affected by the development, however, the relation of per capita costs and per capita benefits seems to be highly undesirable.

Benefits		Costs	
Total	Per capita (gained regionally)	Total	Per capita (incurred locally)
100,000	0.1	10,000	100

Table 1. Distribution of benefits and costs of a hazardous waste facility (Davy 1997)

This theoretical "benefit-cost-distribution" model actually corresponds with experience gained from real planning cases. As Wolsink et al. (2000) notes, "people generally do not come forward with positive responses to planners' agendas". Siting hearings and consultations are forums "where criticism is not only accommodated, it is solicited" (Kahn 2000). Taken wind energy farms as an example, Bell et al. (2005) suggest, that the design of planning processes unintentionally contributes to a bias of public perception of the acceptability of planned facilities. A planning scheme that starts with an initial siting proposal made by facility developers and a subsequent announcement to the public followed by the defense against public criticism provides protest rather than support. Bell et al. suppose a "democratic deficit", taking into account that opinion surveys indicate high support to wind energy use whereas particular wind energy projects often fail due to local opposition.

2.3 Controversial assessment of the NIMBY phenomenon

NIMBY responses have been subject of highly different characterization in the planning and social science literature. Both, negative and positive assessments exist. Some scholars regard

NIMBY and LULU opposition as being motivated by narrow self-interest. Following this characterization, a relatively small group of individuals may effectively put a facility project down ignoring the preferences of the majority. Moreover, NIMBY resistance could have the effect of successfully biasing decision makers' perceptions of community preferences. This effect is intensified by the fact that NIMBY opponents are typically older, more highly educated, wealthier, more likely to be homeowners, and thus more likely to be vocal and politically influential (Schively 2007, p. 257).

Other researchers consider NIMBY opposition - to a certain extent - as a normal form of "grassroots" democracy. NIMBYism stimulates a democratic discourse that ensures a higher quality of siting decisions. Active opposition against proposed development projects sensitizes decision makers and developers to the needs and concerns of affected residents and motivates to implement more sophisticated forms of participation, empowerment and consensus building.

3. Consensus Building Through Procedural Fairness

Social scientists claim that planners, regulatory agencies and politicians have narrowed siting approaches to a task of technical optimization. Such "orthodox" siting, as it is addressed by Davy (1997, p. 3), focuses on four main criteria:

- Profitability: facilities under consideration must yield a benefit to the operator regardless of its status as private or public.
- Functionality: the development of a facility must consider all technical aspects to ensure a functional operation.
- Safety: the development must avoid all harm, risks, and other adverse effects to human health and environment.
- Legality: the facility must meet legal standards.

This traditional approach presupposes - following Davy - that profitable, functional, safe, and legal facilities should be built. However, ensuring that these attributes are met does not necessarily guarantee public support. Based on numerous cases of "informative failures" of facility siting projects, Freudenburg (2004, p. 154) observes an ongoing ignorance of the advice of social research. He claims that planners, regulators and facility managers still ignore perception-related impacts of facilities such as health risk or community stigmatism and demonstrate unprofessional reactions to critics as being emotional, misinformed or irrational.

Owens (2004) warns against simplifying siting controversies to a clash of national or regional interests and local concerns. Following this view, opposition against projects tends to be marginallzed as being subordinated to "higher interests". The proclamation of national needs or the essential importance of projects may cover a lack of real consensus about need and could be a source of mistrust of the "real" interests of projects proponents. "This storyline overlooks the fact that need itself - and conceptions of 'the natural interest' - are often contested; it implies falsely, that issues raised in the form of local inquiry must thereby be 'local' in nature; and it assumes that generic and local considerations can be separated, and dealt with in a neatly hierarchical fashion "(Owens 2004, p. 110). Owens claims that a constructive debate about the desirability of facilities (or land uses) has often been the result of local controversies at the "grassroots" level. Should we aim to meet demand following a traditional "predict and provide" scheme or should we manage demand in another way? The public planning and permit system should encourage such broader, more strategic considerations of planning problems than preventing them.

Another serious source of mistrust is the observable or assumed tendency of locally unwanted land uses to be allocated in socially distressed or "politically weak" areas regardless of technical suitability criteria. As Freudenburg (2004, cited from Owens 2004, p. 104) cynically noticed, "it's funny how technical criteria tend to be satisfied on the poor side of town".

The crucial challenge for successful siting of problematic land uses is consensus building. Incorporating consensus building efforts into siting processes requires more than simple public hearings where "top-down information" on what is planned and the likely effects of the plan is presented. Schively (2007, p. 261) points out, that negotiation is "perceived as the fairest and most acceptable mechanism for siting ... facilities". At the same time, empirical evidence shows that informal processes seem to be more effective in promoting consensus than "official" consultation. Because such forms of communication and negotiation are time-consuming and associated with results difficult to anticipate, facility managers aim to avoid them. However, the likelihood of consensus situations - and actually successful siting - increases with the quality of communication and the perception of procedural fairness by the affected stakeholders. Risk-communication must encompass the full range of stakeholders concerns. Next to technical issues, the study of potentially adverse effects, carried out by the permit agency or the facility proponent, should also address risks associated with reductions in property values and impacts on the quality of life or the image of the host community. Arguments of residents should never be marginalized as irrational, emotional or ignorant to the facts.

Furthermore, a siting procedure and the final decision should be acknowledged as being fair. Fairness demands for an open minded discussion of all benefit and cost factors of the project under consideration. Planners and regulatory agencies are well advised to avoid a "MAD approach to decisions making - to Make, Announce, and then Defend a choice of a preferred site" (Freudenburg 2004, p. 165). If the affected community is confronted with irreversible decisions, a siting approach tends to exacerbate social opposition and reduce trustworthiness of public institutions. Fairness also includes "geographic fairness". As Kunreuther and Susskind (1991) note, it is not fair to locate a critical mass of noxious facilities in a single community or region, even if local residents are willing to accept them. Finally, the aim for fairness could incorporate compensation for the host community or region. In order to limit the intrinsic dilemma of an unequal benefit-cost-distribution with siting decisions, transfer payments to the host community can be an effective means of consensus building. However, compensation may not be successful in cases, when moral or ethical concerns are the key drivers of local opposition.

Based on a national workshop held in 1989, US siting experts issued guidelines for an effective facility siting process (Table 2). The so called "facility siting credo" summarizes the above reflected recommendations of social and political researches and may be used as some kind of checklist for examining the procedural appropriateness of any planning scheme that deals with LULU siting.

Guidelines for an effective facility siting process	
Institute a broad based participatory process	Representatives of all affected groups should be invited to participate in and be assisted at each stage of the siting process
Achieve agreement that the status quo is unacceptable	A siting process must begin with an agreement that a facility is needed. The relevant stakeholders need to understand the consequences of doing nothing
Seek consensus	A serious attempt should be made to involve all the relevant stakeholders to address their values, concerns, potential needs and wants
Work to develop trust	Lack of trust is perhaps the most important barrier to reaching consensus. Those attempting to site a facility must recognize potential sources of mistrust
Choose the solution that best addresses the problem	Problems must be addressed with a facility design and a solution that stakeholders can agree as appropriate
Guarantee that stringent safety standards will be met	No community should be asked to compromise its basic health or safety so that a facility can be built. Preventive measures for reducing the hazard should be encouraged and the proposed facility must meet all health, safety and environmental standards
Make the host community better off	If facilities respond to real needs the magnitude of benefits should be large enough for transfer payments to be made to the host community
Fully address all negative aspects of the facility	When impacts cannot be prevented or mitigated to the satisfaction of the affected parties, various forms of compensation can be negotiated
Use contingent sites through agreements	Some concerns about the management of facilities can be resolved by specifying contingent agreements that spell out what will be done in case of accidents, interruption of service or changes in standards
Seek acceptable sites through a volunteer process	Encourage communities, regions or states to volunteer sites indicating that it is not an irreversible commitment and that there are potential benefit packages (e.g. new revenues, employment, tax reductions) that come with the facility
Consider a competitive siting process	Assuming that multiple acceptable volunteer sites are found, facility sponsors should consider a competitive process of site selection
Work for geographic fairness	It is inappropriate to locate too many noxious facilities in a single locale even if a community is willing to accept them
Set realistic timetables	It is appropriate and helpful to set and enforce realistic deadlines
Keep multiple options open all the time	It is never a good idea to have just one possible site for a LULU even at the final stage of the process

Table 2. "The facility siting credo" (cited from Kunreuther & Susskind 1991)

4. Implementation of Multicriteria Site Selection Methods

4.1 The nature of spatial multicriteria decision making

The siting of noxious facilities or other kinds of locally unwanted land uses is a typical spatial decision problem. It involves a set of geographically defined alternatives, from which a choice is to be made based on a transparent set of evaluation criteria. The decision problem covers five components: (1) a set of goals that represents the normative foundation of the final decision, (2) a set of evaluation criteria or attributes on the basis of which the decision or policy maker evaluates alternatives, (3) a set of geographical alternatives and (4) information regarding the outcome or consequences associated with each alternative (Malczewski 1999).

Multicriteria methods are usually categorized as discrete or continuous, depending on the domain of alternatives. The first approach (called multiattribute decision making) deals with a discrete, limited number of predefined alternatives. The latter (multiobjective decision making) operates with variable decision values to be determined in a continuous domain of a quasi infinite number of feasible alternatives (Malczewski 1999). Table 3 gives an overview on the different nature of both types of multi-criteria decision making.

Criterion	Multiattribute decision making	Multiobjective decision making
Domain	Discrete (pre-specified alternatives)	Infinite (unlimited number of alternatives)
Objectives	"Open" objectives evaluation criteria related to objectives	Explicit goals (e.g. min-/max- targets)
Assessment and Aggregation	Performance table with criteria scores Aggregation of criteria scores with defined rules (preferences as weights)	Use of objective functions with given restrictions Preferences as weights related to different objective functions
Dominant data types in GIS environments	Vector	Raster
Result	Selection of the "best" alternative (ranking)	Calculation of "feasible" alternatives
Well-known methods (examples)	Analytic Hierarchy Process (AHP) Outranking-methods Utility Analysis	Goal programming

Table 3. Multiattribute and multiobjective decision making

Usually, siting approaches combine methods of multiattribute and multiobjective decision making. A "top-down" screening that makes use of Geographic Information Systems (GIS) often uses a set of exclusionary criteria to limit the geographical scope of more detailed analysis of potential sites. Subsequently, a set of specified alternatives are ranked with suitability criteria.

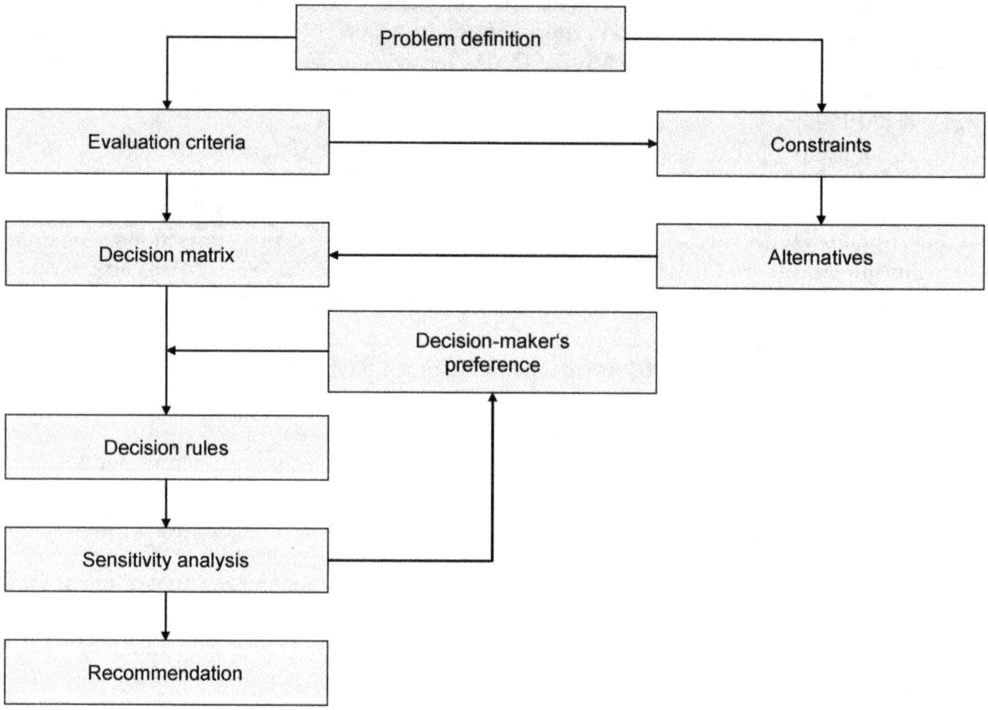

Fig. 1. Framework for spatial multicriteria decision making (adapted from Malczewski 1999)

Figure 1 outlines the basic approach of a multicriteria decision analysis (Malczewski, 1999). After defining the problem - here, a suitable site for a specific land use has to be found within a given normative decision space - certain evaluation criteria are determined. The evaluator should formulate a comprehensive set of objectives that reflects all concerns relevant to the decision problem and objective-related measures (attributes). After excluding areas that are considered as non-suitable by using constraint criteria, a set of feasible alternatives describe the decision space. A constraint represents natural or political/social restrictions on the potential alternatives. Constraint analysis is usually carried out with conjunctive and disjunctive screening methods or with the use of target constraints (e.g. as demanded minimum or maximum attribute values). Subsequently, the specified alternatives are described with a decision matrix that displays all attributes of all alternatives (also called performance table). The normative basis of the final evaluation is the degree to which the objectives are fulfilled, measured by attributes. However, at least two methodological steps have to be made before ranking the alternatives according to their objective-related performance. First, the process of ranking decision alternatives typically involves criteria of different importance to the decision situation. The evaluator or decision maker has to assign weight factors to each criterion. Weights indicate the relative importance of objectives or attributes to other criteria under consideration. Secondly, the alternative's attributes are normally measured in different scales, whereas most multicriteria methods require that attributes are expressed in a similar scale. Therefore, a standardization procedure has to be carried out (e.g. a linear transformation procedure.

With a given set of alternatives, the decision matrix with standardized attributes, and the predefined weights, the final decision is just a formal step. The decision maker has to select a decision rule that provides an ordering of all alternatives according to their objective-related performance. What kind of decision rules is considered as appropriate depends on the specific decision situation. In many cases, a simple additive weighting will meet the requirements; additive weighting means to calculate a total score for each alternative by multiplying the (standardized) attribute values by their weight factors and summing the products over all attributes. The alternative with the highest individual score is regarded as the preferred one. However, this kind of aggregation is restricted to decision situations where linearity of attributes can be assumed. Linearity means that the desirability of an additional unit of an attribute (e.g. hectare, kilometer, individuals) is constant for any level of that attribute.

Alternatively, the decision maker can use value/utility function methods to aggregate the attribute values for the final decision. Here, attribute utility functions are used to transform attribute values into an interval-utility scale (compare the comments to Utility Analysis).

What is the contribution of multicriteria decision methods in coping with siting conflicts? Methods like the Analytic Hierarchical Process (AHP) or even simple computer-assisted overlay mapping techniques can help to overcome opposition by supporting a transparent, trustful planning process. Transparency of information (data sources and indicators used) and normative assumptions (e.g. criteria weights) is a prerequisite of effective communicating about risks of planned facilities. In contrast to a solely "political" decision based on a set of qualitative expressions of preferences and a non-quantitative aggregation, formal multicriteria decision methods allow critics to "decompose" the decision for a site (or the exclusion of alternatives) in every detail.

Of course, "top-down" siting procedures should be reflected against bottom-up considerations derived from local hearings or more sophisticated forms of consensus building. Freudenburg (2004, p. 157) strongly recommends the incorporation of local knowledge into technical site selection approaches: "The problems with [...] top-down approaches often become more evident, for example, in the face of the fact that local citizens may know more about certain characteristics of local sites than will be available in the aggregated data used by the GIS analyst, leading to conversations along the lines of, 'If this is supposed to be a scientific process, how could you have "overlooked" something that everyone [here] knows?'" The results of multicriteria analysis should never be presented as the ultimate technical solution of a decision problem which makes any kind of further consideration needless. Therefore, quantitative multicriteria decision techniques following a rational and logical planning credo on the one hand and forms of local negotiation and consensus building on the other hand are complementary not exclusionary.

4.2 Overview on multicriteria analysis methods

There are numerous methods for structuring a decision problem, evaluating feasible alternatives and prioritizing alternative decisions that can be implemented in siting procedures (see Malczewski 1999 and Malczewski 2006 for an overview on methods). In this subchapter, only some of them will be briefly described.

4.2.1 GIS-based overlay mapping

Overlay mapping is one of the most frequently used methods in environmental planning. Its basis approach is relatively simple. Following a given problem definition, certain evaluation criteria resp. attributes are presented in the form op maps or map layers in a GIS environment. Each map can be regarded as an individual suitability map with respect to the land use under consideration. Based on defined aggregation rules (see above), these maps will then be combined to provide an overall suitability map. GIS software provides the operator with a broad range of tools related to map algebra techniques. Therefore, if appropriate geodata sources are available, overlay mapping is quite easy to implement.

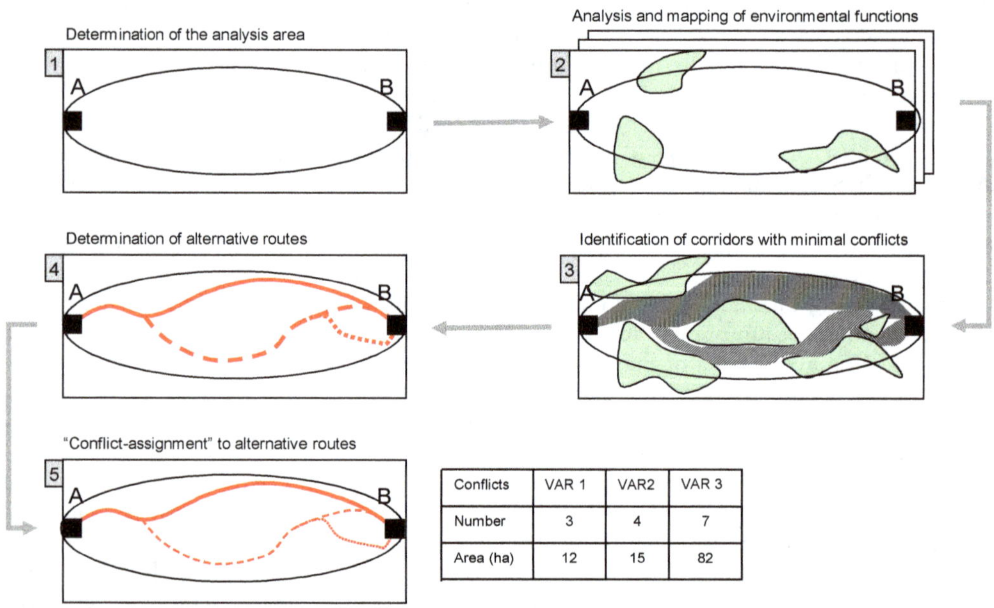

Conflicts	VAR 1	VAR2	VAR 3
Number	3	4	7
Area (ha)	12	15	82

Fig. 2. GIS-based identification of infrastructure corridors with minimal environmental conflicts

Figure 2 shows the workflow of an overlay mapping approach used in transport planning in Germany. The procedure intends to identify a suitable corridor for a road or railway track in an early stage of planning. The "suitability" of potential corridors is assessed by their potential conflicts with environmental and social values. After determining the study area (phase 1), environmental and social values that might indicate natural or social constraints for infrastructure planning (e.g. protected habitats that might be dissected or sensitive urban functions that are affected by noise emissions) have to be mapped and organized in a GIS layer structure (phase 2). Based on a spatial overlay of potential constraints and conflicts, alternative corridors with an expected minimum number of conflicts are determined (phase 3 and 4). Finally, all alternatives are compared with respect to their conflict intensity (phase 5). A simple summation of function-specific conflicts can be used here.

Another overlay mapping method, popular in German environmental planning, is called Ecological Risk Assessment (ERA). The method attempts to estimate the "ecological risk" of projects in situations that are characterized by a high degree of uncertainty. In ERA, "risk" means the possibility of threats to valued natural assets and ecological components. The estimated risk is regarded as the product of natural vulnerability and the level of perturbation (or disturbance) due to the project under consideration. Risk modeling in ERA follows the common rule that the higher the vulnerability and the level of perturbation, the higher the risk of an environmental damage.

The method is organized in three steps. In step 1, the potentially affected area by the project and its physical features has to be analyzed. Step 2 attempts to assess the level of vulnerability based on a thorough analysis of valued ecological components (or functions). The results of this analysis are stored as a series of GIS layers. With step 3, the ecological risk has to be estimated. Usually, a simple matrix with ordinal scales for addressing vulnerability and perturbation features is used for this final step (Figure 3). Map algebra functions technically support this kind of risk modeling in a GIS environment.

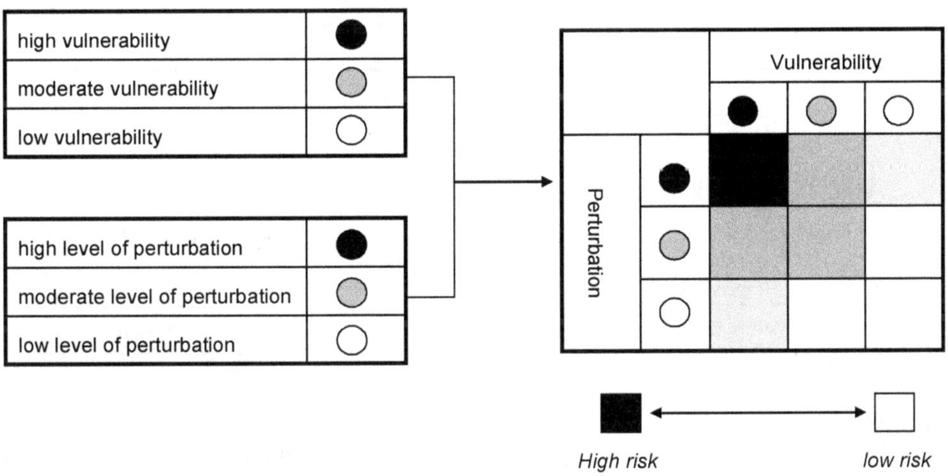

Fig. 3. Risk-assessment scheme in Ecological Risk Assessment (ERA)

4.2.2 Analytical hierarchical process

The Analytic Hierarchical Process (AHP) - developed by Thomas Saaty in 1980 (Saaty 1980) - requires the operator to decompose a decision problem in form of a hierarchy of objectives, criteria and alternatives (Figure 4). The method involves one-on-one comparisons between each element of a certain hierarchy level. Pairwise comparisons are used to assign relative weights on the objectives and criteria based on a standard ratio scale (Table 4). Saaty introduced different approaches to calculating relative weights based on a pairwise comparison matrix. The result is a composite set of priorities for the lowest tier of the hierarchy, namely the alternatives.

One of the main advantages of the method is the fact that it is able to process information of different scales. Qualitative judgements ("A is much more important than B") are handled in the same way as numeric values ("A is 5.4 whereas B is only 2.9").

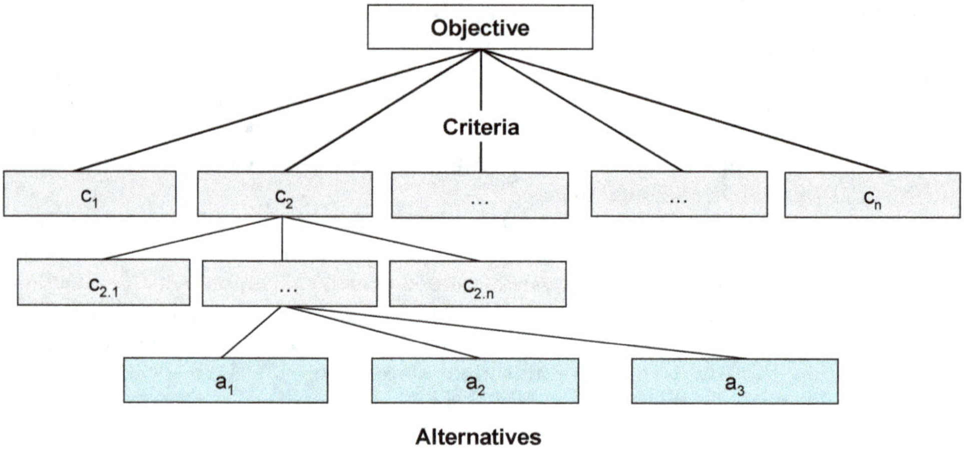

Fig. 4. Hierarchical structure of a decision problem within the AHP process

Scale	Meaning/Interpretation
1	same
3 (1/3)	a little bit larger (smaller) or more important (less important)
5 (1/5)	significant larger (smaller) or more important (less important)
7 (1/7)	much larger (smaller) or more important (less important)
9 (1/9)	very much larger (smaller) or more important (less important)

Table 4. The AHP standard scale for pairwise comparisons

4.2.3 Utility analysis

Next to Cost-Benefit Analysis, Utility Analysis (UA) is one of the best-known multicriteria analysis methods used in environmental and infrastructure planning in Germany (see Figure 5). The key principle of UA approaches is the transformation of attribute values of different scales into an interval (value) scale, usually a standard scale ranging from 0 to 100 or 0 to 1.0. The transformation process requires criteria-specific transformation functions (also called utility functions), which reflect the decision maker's preferences. The transformed values are aggregated into a total utility value that represents the performance of an alternative. Weights are used to express the different importance of the employed criteria. The multiplication of (criteria resp. attribute specific) utility values by the determined weights leads to partial utility values. In the standard procedure of UA, the final aggregation is carried out as a simple summation of partial utility values. The alternative with the highest total utility value is the preferred one.

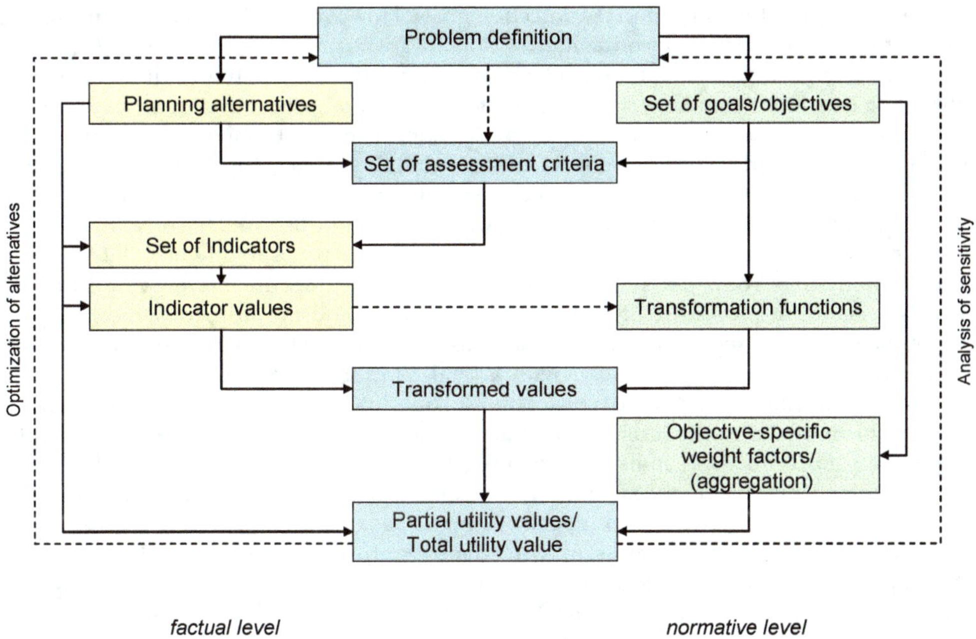

Fig. 5. Basic scheme of Utility Analysis methods (adapted from Bechmann, 1989)

It should be emphasized that UA approaches underlie one crucial assumption: the additivity of attributes. The additivity assumption requires that there are no interaction effects between the selected attributes. Complementarities between attributes may lead to inappropriate results. Therefore, the implementation of UA methods should be based on a thoroughly carried out theoretical analysis of the decision situation.

4.3 Case study: the siting of wind energy farms in Germany

Due to massive public funding, Germany experienced a tremendous growth in wind energy production in recent years. Currently, more than 18,000 wind energy plants with a capacity of 20,000 MW are installed throughout the country with spatial hubs in coastal and "flat" regions of the North. In 2006, the share of wind energy to total electricity consumption was more that 6%. Like in other western countries, wind energy planning in Germany is characterized by high public support of wind energy use in general but massive opposition against local windfarm projects.

After experiencing a phase of chaotic spread of wind mills in the 1990s, the German legislator adopted some amendments to federal regional and urban planning codes in order to achieve a more controlled wind energy planning. Henceforward, the use of wind energy outside urbanized areas ("Außenbereich") was regarded as privileged. "Privileged" means that certain kinds of land uses are permitted in general without making any arrangements for their location. Developers must get permission unless public concerns are opposed to a specific (privileged) land use. Taken wind energy use as an example, relevant concerns could encompass negative effects to scenic values, threats to well-being of residents nearby proposed mills or nature and species protection goals. However, the legal barriers for permit agencies to deny permission are quite high.

At the same time, regional and local planning administration got the right to effectively manage the location of wind energy mills by means of spatial concentration zones as well as "no-go" zones for future wind energy production. The most powerful instrument of regional and local land use planning is called suitability area ("Eignungsgebiet") where specified land uses (e.g. wind mills) are to be concentrated (see § 7 Sec. 4 No. 3 of the Federal Regional Planning Act). Within the suitability area, the land use under consideration has priority against rivaling land uses. Outside the area, the land use is totally prohibited.

Based on numerous court decisions and planning guidance documents provided by state agencies, a standard procedure of wind energy planning (and the siting of wind mills) has been implemented in regional and local land use planning. Most importantly, the courts consider negative planning associated with a total ban for privileged land uses as illegal. The Federal Administration Court has pointed out that the exclusion of wind energy production from parts of the jurisdiction is justifiable only in cases when the land use plan secures the priority of wind mills against other land uses on other suitable lands. Simply spoken, a community that dislikes wind mills is not allowed to ban them from their territory by exclusionary zoning. German courts demand a coherent planning concept that acknowledges the privileged status of wind energy production outside urbanized areas without violating the legal rights of other land users. Therefore, an area-wide and integrated suitability analysis is regarded as crucial to meet the legal requirements for wind energy planning.

The suitability analysis is usually organized as follows:

- In step 1, areas that are regarded as non-suitable for wind mills are excluded from further analysis; Table 5 outlines a set of exemplary criteria for the exclusion of "no-go areas".
- In step 2, areas with wind speeds below commercial standards have to be excluded from further analysis.
- Step 3 aims to model the conflict potential in the remaining areas after excluding no-go areas and areas with unsuitable resource quality. For this purpose, a set of criteria indicating conflicts with other land uses is used. Areas with a critical spatial overlay of conflicts are excluded. Often, a simple additive weighting is used to determine those areas.
- Step 4 excludes smaller areas below a threshold value (e.g. 20 hectares) to avoid a spatial dispersion of small wind farms. However, the relevance of step 4 depends on whether regional or local policy makers prefer a lower number of larger wind farms (with more than 10 or 20 mills).
- Finally, step 5 undertakes an individual assessment of remaining areas with technical and economic criteria (e.g. accessibility by road or tracks, connectivity to existing power lines) as well as small-scale conflict criteria (e.g. soil features, distance to farms or small settlements).

This stepwise suitability analysis can be effectively supported by GIS tools. Both, raster and vector data analysis will be relevant for solving the siting task.

Criterion	Value
Distance to urbanized areas	< 700 m
Distance to four-lane motorways	< 40 m
Distance to two-lane federal and state roads	< 20 m
Distance to railway tracks	< 50 m
Nature protection areas	Area with a 200 m buffer
Nature protection areas of European importance (FFH and bird protection areas)	Area with a 1.000 m buffer
Distance to rivers and creeks	< 10 m
Protected forest areas	Area with a 200 m buffer
Areas for groundwater protection	Area

Table 5. "No-go areas" for wind mill siting in Baden-Württemberg

5. Conclusion

The NIMBY syndrome is by no means an impregnable barrier towards successful facility planning. The way in which planners and engineers deal with NIMBY attitudes held by local residents highly influences the viability of resistance and the outcome of planning. Planners should learn from "informative failures" and improve the quality of procedural standards. Procedural fairness, based on a broad risk-communication, is a crucial prerequisite in successfully coping with NIMBY opposition. GIS-based multicriteria analysis methods may help to slow down protest by supporting a transparent, trustful planning process. Providing transparency of information and explicit or implicit normative assumptions is an effective means of communicating about risks of planned facilities. It should be emphasized that quantitative multicriteria decision techniques, following a rational and logical planning credo, on the one hand and forms of local negotiation and consensus building on the other hand are complementary not exclusionary.

6. References

Bechmann, A. (1989). Die Nutzwertanalyse, In: *Handbuch der UVP,* Storm, P.C.; Bunge, T. (Ed.), 31 p., Abschnitt 3555, Berlin

Bell, D.; Gray, T. & Haggett, C. (2005). The 'social gap' in wind farm siting decisions: explanations and policy responses. *Environmental Politics,* Vol. 14, No. 4, 460-477

Davy, B. (1997). *Essential Injustice - When Legal Institutions Cannot Resolve Environmental and Land Use Disputes,* Springer, New York

Fischel, W.A. (2001). Why are there NIMBYs?. *Land Economics,* Vol. 77, No. 1, 144-152

Freudenburg, W.R. (2004). Can we learn from failure? Examining US experiences with nuclear repository siting. *Journal of Risk Research,* Vol. 7, No. 2, 153-169

Kahn, R. (2000). Siting struggles: the unique challenge of permitting renewable energy power plants. *Electricity Journal*, Vol. 13, No. 2, 21-33

Kunreuther, H. & Susskind, L.E. (1991). *The Facility Siting Credo: Guidelines for an Effective Facility Siting Process*, Publication Services, University of Pennsylvania, Philadelphia, PA

Lober, D.J. (1995). Why protest? Public behavioural and attitudinal response to siting a waste disposal facility. *Policy Studies Journal*, Vol. 23, No. 3, 499-518

Malczewski, J. (1999). Spatial multicriteria decision making, In: *Spatial Multicriteria Decision Making and Analysis. A Geographic Information Sciences Approach*, Thill, J.-C. (Ed.), 11-48, Aldershot et al., Ashgate

Malczewski, J. (2006). GIS-based multicriteria decision analysis: a survey of the literature. *International Journal of Geographical Information Science*, Vol. 20, No. 7, 703-726

Owens, S. (2004). Siting, sustainable development and social priorities. *Journal of Risk Research*, Vol. 7, No. 2, 101-114

Saaty, T. (1980). *The Analytic Hierarchy Process*, McGraw-Hill, New York

Schively, C. (2007). Understanding the NIMBY and LULU phenomena. Reassessing our knowlegde base and informing future research. *Journal of Planning Literature*, Vol. 21, No. 3, 255-266

Wolsink, M. (1994). Entanglement of interests and motives: assumptions behind the NIMBY-theory on facility siting. *Urban Studies*, Vol. 31, No. 6, 851-866

8

Partnership between Municipality and Public University to Improve the Sustainable Development of Small Municipalities

Camilo Michalka Jr.
Federal University of Rio de Janeiro (UFRJ) - Urban Engineering Program
michalka@ufrj.br
Brazil

1. Introduction

Rio de Janeiro suffers from the fast growth of migration, which is approaching the city from the smaller municipalities and rural areas. This causes a variety of tremendous social problems. One of them is the lack of a complex social housing program to manage the highly needed residential space for the new inhabitants. In the same time, it makes the cities aim to provide a sustainable urban development of the Metropolitan Area of Rio de Janeiro a lot more difficult. The migration in the metropolitan areas has been very intensive in Brazil along decades. In the history of these metropolises, the necessary planning for technical and social infrastructure to receive this migration was missing.

Another problem is that the most part of the people that migrate to the big cities, do not have a profession. They look for works that do not need qualification. Because of this, they do not have a good salary, which would permit support their homes. As consequence appear the slums (Safatke & Bernardi, 2008).

One of the solutions for this urban situation is to establish and improve the render and substantial improvement of the smaller cities' development and education, which are responsible for the migration towards the Metropolitan Areas.

The tasks for the smaller cities are to develop sustainable concepts as an answer to the huge pressure on the migration into the Metropolises, in order to integrate the city in the countries' necessities and further on to reduce the migration from the small cities towards the big cities and metropolitan areas.

In this concept, it is necessary to improve a professional education, to give the inhabitants opportunity to work in professions that are necessary to the economical reality of the municipality. The most part of the young people do not have yet opportunities to learn a profession in his municipality.

It is important to consider that the small cities in Brazil, still have the opportunity to implement a sustainable development and growth of their own municipality. The scale of these cities can still be influenced and has several chances to establish a planning structure. This focus can contribute to solve these tasks of an accelerating agglomeration of inhabitants

that produces tremendous areas of informal settlements like in the Metropolitan Area of Rio de Janeiro.
It happens frequently that Brazilian and international efforts are combined to solve problems in the Metropolis, like in the urban area of Rio de Janeiro.

A project from *Escola Politécnica* of Federal University of Rio de Janeiro, is to work with this new perspective, which means to handle the Metropolis, by looking at one of the most important origins of the problem, as there is the migration towards the Metropolises.
To this propose it was taken the municipality of São José do Vale do Rio Preto, in the state of Rio de Janeiro (ca. 20,000 inhabitants) as a study case (Michalka, 2007b).
The work is focused on the prior action lines:

- master plan;
- land use;
- technical infrastructure;
- social infrastructure;
- environmental planning;
- local sustainable development;
- regional development.

The projects aim is to provide methodologies, concepts and ideas for the municipality so that it will be able to develop and establish a sustainable planning structure for its municipality independently. The exchange of ideas about the reality of the community and the look for a development with the contributions of experts is a pilot project for the city as well as for the academy, and an important step towards a sustainable future.
The focus will be on Brazil, but it is easily adaptable to many developing countries.
One of the objectives of the pilot project is, to give the academy an overview of the big differences between small and big communities. The aim is to make both aware of what happens to most of the people who move from a small community to a megacity, often in order to improve their life.
The small city that was chosen for this work is *São José do Vale do Rio Preto,* located in the state of Rio de Janeiro, 110 km north of the City Rio de Janeiro. The *Escola Politécnica* of Federal University of Rio de Janeiro works in this project with the local authorities since the year 2000. The University of Applied Sciences Berlin is also participating.

1.1 Geographical and Political Context of Brazil

Brazil covers 8,500,000 km^2 with a population of 188,300,000 people in 5,564 municipalities. That makes 22.11 hab/km^2, which means that, there is a lot of space for each individual.
Of course the Brazilians don't spread out equally over the country - not even in regions with similar geographical characteristics. Figure 1 shows the political division of Brazil, with the 26 states with their capital and the Government District.
Table 1 shows the distribution of the population in these 26 States of Brazil and the Government District, as well as the rate of inhabitants per square kilometer in percent as well and the number of municipalities in each state.

State	Capital	Number of Municipalities	Area (km^2)	Population (Estimation 2006)	Inhabitants per km^2
Acre	Rio Branco	22	152,581	686,652	4.50
Alagoas	Maceió	102	27,767	3,050,652	109.86
Amapá	Macapá	16	142,814	615,715	4.31
Amazonas	Manaus	62	1,570,745	3,311,026	2.10
Bahia	Salvador	417	564,692	13,950,146	24.70
Ceará	Fortaleza	184	148,825	8,217,085	55.21
Distrito Federal	Brasília	1	5,802	2,383,784	410.85
Espírito Santo	Vitória	78	46,077	3,464,285	75.18
Goiás	Goiânia	246	340,086	5,730,753	16.85
Maranhão	São Luís	217	331,983	6,184,538	18.62
Mato Grosso	Cuiabá	141	903,358	2,856,999	3.16
Mato Grosso do Sul	Campo Grande	78	357,125	2,297,981	6.43
Minas Gerais	Belo Horizonte	853	586,528	19,479,356	33.21
Pará	Belém	143	1,247,689	7,110,465	5.69
Paraíba	João Pessoa	223	56,439	3,623,215	64.19
Paraná	Curitiba	399	199,314	10,387,378	52.11
Pernambuco	Recife	185	98,311	8,502,603	86.48
Piauí	Teresina	223	251,529	3,036,290	12.07
Rio de Janeiro	Rio de Janeiro	92	43,696	15,561,720	356.13
Rio Grande do Norte	Natal	167	52,796	3,043,760	57.65
Rio Grande do Sul	Porto Alegre	496	281,748	10,963,219	38.91
Rondônia	Porto Velho	52	237,576	1,562,417	6.57
Roraima	Boa Vista	15	224,299	403,344	1.79
Santa Catarina	Florianópolis	293	95,346	5,958,266	62.49
São Paulo	São Paulo	645	248,209	41,055,734	165.40
Sergipe	Aracajú	75	21,910	2,000,738	91.31
Tocantins	Palmas	139	277,621	1,332,441	4.79
BRAZIL	**Brasília**	**5.564**	**8,514,205**	**188,298,099**	**22.11**

Table 1. Distribution of the population in the states of Brazil (www.portalbrasil.com.br)

Fig. 1. Political division of Brazil (www.portalbrasil.net/brasil)

The urbanizing process in Brazil has increased strongly, which can be seen in Table 2. We verify that in 1960 there were 45% of the population in urban areas, and in the following decades it has been increased to 55.9%, 67.6%, 78.3% and reached 81% in year 2000.

Not only the migration of people towards the cities that causes this enormous urbanization growth. It started out with the migration to the capitals of the states. In consequence the capitals became metropolitan areas. Table 3 lists the most important metropolitan areas of Brazil and their population.

Population of Brazil							
Total	2000	1996	1991	1980	1970	1960	1950
Urban	137,953,959	123,076,831	110,990,990	80,437,327	52,097,260	32,004,817	18,782,891
Rural	31,845,211	33,993,332	35,834,485	38,573,725	41,037,586	38,987,526	33,161,506
Percentage							
Urban	81.25	78.36	75.59	67.59	55.94	45.08	36.16
Rural	18.75	21.64	24.41	32.41	44.06	54.92	63.84
Reference: Demographic Census 2006							

Table 2. Urban and rural population

Metropolis	Population / 98 - 1000
São Paulo	17,054,900
Rio de Janeiro	12,270,100
Belo Horizonte	3,957,700
Porto Alegre	3,337,500
Recife	3,157,100
Salvador	2,797,600
Curitiba	2,584,900
Fortaleza	2,699,500
Belém	1,550,000
Total - Metropolitan Areas	**47,485,200**
Total - Brazil	**161,340,000**

Table 3. Population in metropolitan areas - Data IBGE 2000

There are 5,564 municipalities in Brazil. In comparison to the metropolises the number of inhabitants in smaller communities varies a lot. Figure 2 shows the amount and size of communities in relation to the number of citizens. Only 0.6% of the cities are megacities, where as 71% of the cities don't have more than 20,000 inhabitants.

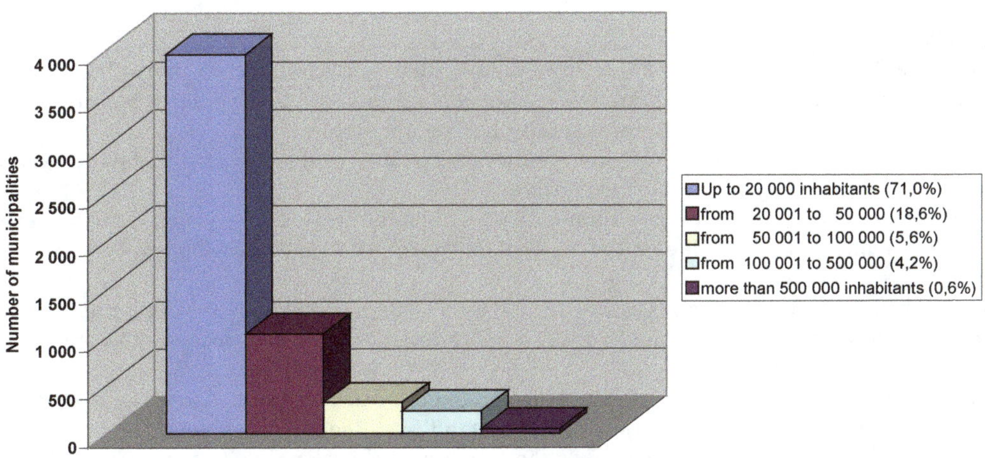

Fig. 2. Distribution of the population in the 5,564 municipalities (www.ibge.gov.br)

2. Considerations about the Urban Development

2.1 Metropolitan Areas

The sustainable urban development needs planning. On the other hand, to plan a sustainable development, it is necessary to consider the influence of each parameter that is present in the nature and how they interact. A systemic approach how explained in Capra (1997). To a sustainable urban development, it is necessary to consider the necessary technical and social infrastructure needs for the process. Its is necessary to take in account haw the different parts of each kind of the infrastructure (traffic, water supply, education among others) interacts.

Most of the urbanization and the urban plans in Brazil haven't considered the projects that included the technical infrastructure and the social infrastructure.

The migration in the metropolitan areas has been very intensive in Brazil along decades. In the history of these metropolises, the necessary planning to receive this migration was missing. Because of that, the enlargement of the cities happened without the necessary infrastructure plans for the migrations process. In the part of the cities where an infrastructure exists, it has been permitted a concentration of inhabitants without considering the impact in the existing infrastructure. Because of this appear the problems with traffic, water, waste, air pollution, sound pollution, health, education and others.

Without a plan to the habitation, the people construct their homes. To this, the people use frequently empty areas, where there is no infrastructure. Appears then the slums. They grow in a spontaneous way. It makes very difficult and expensive to act in order to organize all the necessary urban services. One big problem of the metropolitan areas in Brazil is the big amount of slums, and they are in a permanent grown.

An important consequence of the lack of plan and the existence of the slums in the metropolitan area is that the environment is strongly attacked. The effect of this in the quality of life of the city and the inhabitants is enormous. The humanity is part of the environment and suffers influence from it.

2.2 Small Municipalities

In figure 2 can be seen, that 71% of the Brazil's municipalities have no more than 20,000 inhabitants. In these municipalities, even with existing poverty, there are not big problems with slums. The environment in these small cities is more natural.

It make possible to plan. An intervention to solve problems of inhabitation and infrastructure is easier and not too expensive.

In these municipalities with low number of inhabitants is possible to plan a sustainable development. Taking the title of the book of McHarg (1992) it is possible to *Design with Nature.*

To reach this objective, it is necessary to use the existing knowledge about planning with consideration of the environment, the social necessities and the necessary economical development.

Important tools to plan, for example, the use of geo information systems to make diagnosis, canaries and prognosis are not know or understand in these municipalities. The university possesses a lot of the necessary knowledge to support the planning. An example is in da Silva (2004). The universities have not only the knowledge. Has the tools to apply this information to the particularities of each municipality too.

It can be assert, that the biggest difficulty of the small municipalities is exactly, how to get the necessary information, particularly the technical information, to support the necessary political decisions.

This is the objective of this work, which will be detailed next.

3. The Focus

After this brief introduction, the question of most interest is: which circumstances lead to this situation?

Together with the University of Applied Sciences of Berlin, the Federal University of Rio de Janeiro is working continually with the municipality of São José do Vale do Rio Preto, in the state of Rio de Janeiro.

The state of Rio de Janeiro covers 43,700,000 km^2 with a population of 15,380,000 inhabitants, counted in 2005. 95% of the people are living in urban areas. The state of Rio de Janeiro consists of 92 communities. Figure 3 shows the state of Rio de Janeiro and its political division.

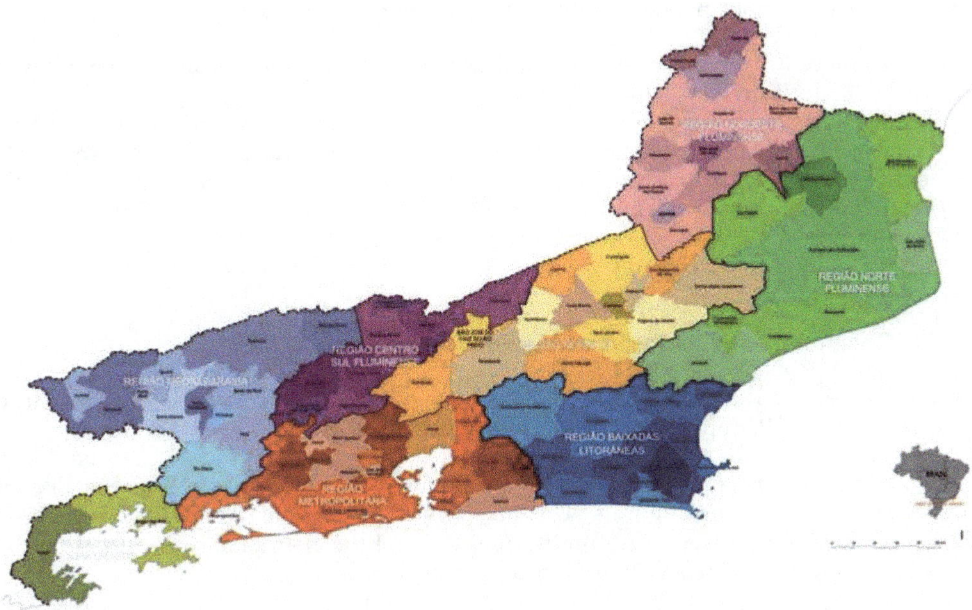

Fig. 3. Political division of the state Rio de Janeiro - 92 communities

This work assumes that there are equal problems in other small communities as the ones we have analyzed in São José do Vale do Rio Preto and considered to be keys for changing the situation. The most important are as follows.
The small communities:

- don't have the support of a skilled technical team in their administrations;
- have difficulties in accessing the needed information;
- have, as a consequence, difficulties in making the best decision, because they neither possess the information nor the staff to analyze and solve the problem.

Dealing with this kind of problems, it is difficult for the small communities to plan the economically growth of the municipality. But in order to offer further jobs in the community, this plans above a must be considered. Lacking this plan means, that on the long run young people won't have opportunities to make their livings in their hometowns. As a consequence these young people migrate to the metropolis looking for jobs and a better future.

Without any doubt the problems of the Metropolises won't be solved, if the small cities don't find solutions for their own people. Of course the Metropolises are trying to improve the situation as well. But if there are no jobs in the small cities meanwhile the big cities seams to provide them - no matter how bad they pay, and how bad the living conditions will be - more and more people will migrate to the big cities. The work of improving a big city will then be without an end and the aims will never be reached.

Nowadays there is a focus on research and necessary projects in the small communities. Therefore the decision was made; to work in cooperation with the community of São José do Vale do Rio Preto, how mention above. The University and the Community have experience in working together and know how to benefit from each other. The Community needs consulting and the University needs to find a way of providing the necessary information and knowledge to the Community in a way that it can be understood and used also by less educated workers. The University needs to take its knowledge "outside its own walls".

3.1 The Community of São José do Vale do Rio Preto

São Jose do Vale do Rio Preto is located in the state of Rio de Janeiro and used to be part of the municipality of Petrópolis. In 1989 it became independent.

São José do Vale do Rio Preto is located in the micro region Serrana (Mountain Region) marked red, in the map of figure 4. This area is well known for its mild climate and the beauty of its landscape, reaching from the mountains of the Natural Park Serra dos Orgãos to the remaining parts of the Mata Atlântica (Atlantic Forest).

Sao Jose do Vale do Rio Preto is in a 110 km distance to the city of Rio de Janeiro which are a 2 hour car drive. It is accessed over the federal road BR 040 on the way from Rio de Janeiro to Belo Horizonte. The other option is the federal road BR116 from Rio de Janeiro leading northeast through the country. Other main cities in reach are Belo Horizonte (370km) and São Paulo (530km).

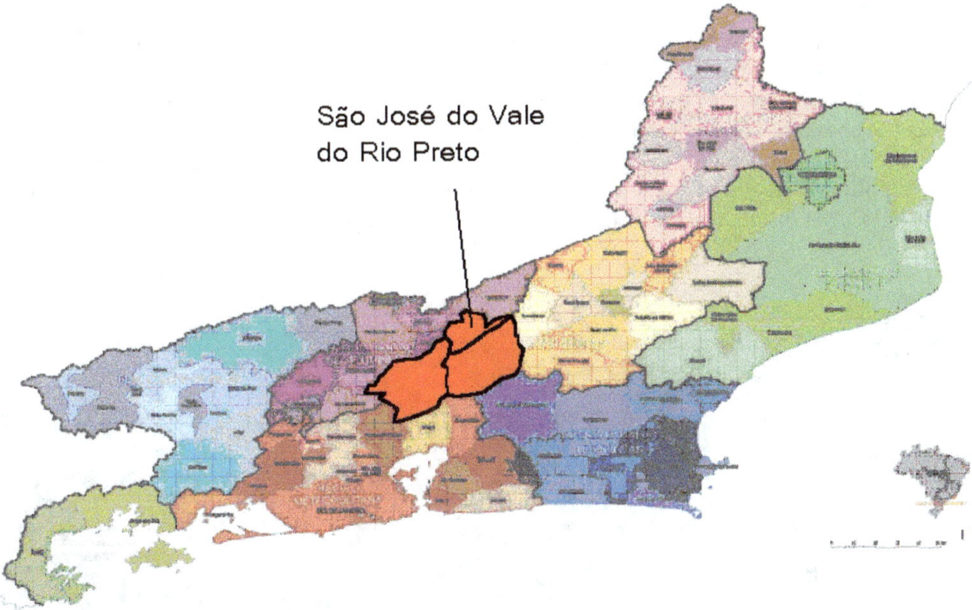

Fig. 4. Mountain region and São José do Vale do Rio Preto

3.1.1 A Short Introduction to São José do Vale do Rio Preto

São José do Vale do Rio Preto is spread out over more 20 km alongside the beds of the river Preto, covering 250 km². On an average height of 600m above sea level, it is surrounded by several mountain ranges. The southern peaks top up to 1,450 m, the northern summits reach 950 m. Sao Jose is split into several smaller communities which form the city with a total of 20,000 inhabitants. Figure 5 shows a overview from São José do Vale do Rio Preto's downtown.

Fig. 5. São José do Vale do Rio Preto downtown

3.1.2 The river

The Rio Preto ("Black River") gains its name from the black stones covering the river bed which is only 1 to 3 meters deep (figure 6). It flows from the municipal of Friburgo in the east to the Paraiba river in the west, which leads into the sea at São João da Barra. The Preto river is the characterizing natural wealth of Sao José.
Unfortunately today, the waters are polluted. When it arrives in Sao Jose the waters are already polluted due to the direct pouring of sewage in the city of Teresópolis. Neither of the cities located on the river possess over a water cleaning system.

Fig. 6. The Preto river

3.1.3 Water supply

The city gains its fresh water basically from two main springs Maravilha tank (60m^3, daily production of 500,000 liters), and Glória (180m^3, daily production of 115,000 liters). This covers 70% of the populations needs. All the distributed water is chlorinated
There are dozens of small gains, that the inhabitants have constructed along side the small rivers to supply their households. In the communities of Barrinha, Pedras Brancas and Contendas, three bigger reservoirs have been built, supplying about 500 inhabitants each.

3.1.4 Population and economy

There is no visible misery in the population of Sao Jose: there are no beggars found in the streets, but none the less, poverty exists.

The poorest part of Sao Jose is the quarter of Pedras Brancas. Recently there are a range of programs in order to improve the situation.

Between 1991 and 2000 the city's average growth rate was 2.5 % per year. The majority of the population is 20 to 49 years old. Nowadays there is no population growth.

In 1950 São José de Vale do Rio Preto was considered the biggest poultry raising center in South America. As a side effect of the poultry farming São José produced tons of natural fertilizers - a contribute to the vegetable production of the following years. Until today São José has the biggest poultry farmer concentration in Brazil.

The agriculture is also important for the city of Rio de Janeiro. The production of São José do Vale do Rio Preto goes to there, so that São José do Vale do Rio Preto is economically dependent of the city of Rio de Janeiro.

3.2 The Work between Federal University of Rio de Janeiro (UFRJ) and São José do Vale do Rio Preto

The associated work of the UFRJ and the city of São José do Vale do Rio Preto is the result of the starting team work of the UFRJ and the University of Applied Sciences Berlin.

The first steps were made in a Seminar in 1999 in Berlin, where Prof. Michalka was present. In 2000, the *Escola Politécnica* of UFRJ invited professors of the Technische Fachhochschule Berlin (TFH Berlin) to a conference in Rio de Janeiro intending to start a team work to do research and initiate projects to develop the environment in small communities. Since then the two universities have been realizing projects together successfully.

An important part of this work is the exchange of students. Thanks to this association, German students have the opportunity to live and work in São José do Vale do Rio Preto and participate in projects, and Brazilian students have the opportunity to study and to make projects in Berlin.

As mentioned above, small communities are dealing with big difficulties in order to access basic knowledge to make wise economical, social and environmental decisions. The participation of students gave the opportunity of providing the community in a simple comprehensive way with the information that is needed. In an exchange the students learned about the "every day problems" a community is handling, how to solve problems and how to develop plans for economical, social and environmental growth. In exchange, the University and their students learn about the process of making decisions in a small city. Many decisions made by the community since the associated working has started, have been influenced by students as well as by meetings with professors from the UFRJ and some visits of professors from the TFH Berlin in São José do Vale do Rio Preto. Since 2002 there are regular meetings between the *Escola Politécnica* of UFRJ and authorities of the municipality.

Between 2005 and 2007, the UFRJ took part in the ALFA Project (project ALFA AURORA). This has allowed that Students from Holland, Spain and Italy have also worked in São José do Vale do Rio Preto (Michalka, 2007b).

It is important to stand out that the university do not want to take the decisions on what the municipality must do. The objective of the contribution of the University to the Municipality is to give the necessary knowledge to permit the municipality can take de best political

decision. Is to permit that decision can be based on technical information. The university can also help to create scenarios to make prognoses of the consequences of different decisions. Results of the associated work can be seen in Michalka (2007a) and Tem Hold et al. (2007).
As example, two initiatives from the long list of successful teamwork results were chosen which have become real in the municipality.
The first one deals with sewage. With a student concept for a sewage system, the community was able to initialize a project and also gained money to start constructing a small part of the sewage net. Because of the convincing results the community won more money to be able to do the same work it in city center as well.
The second example is in the public houses of the executive, legislative and judicative powers. Taking in account that the community only exists since 1989, it is not astonishing, that the three powers are jet not housed efficiently. The community has done research with the result that it would be best to construct new buildings. Also because the necessary space for this project could easily be provided. In the first place it was planned to construct only the public buildings. Then students made further suggestions to improve the concept. As a result today's concept is to build a whole new quarter that covers the needs of a lot of different interest such as public housing, trade and habitations, with the actual knowledge for urban planning.
Because the good results in the partnership between the *Escola Politécnica* of UFRJ and São José do Vale do Rio Preto, the *Escola Politécnica* intents to enlarge and deepen the partnership. One objective is being discussing nowadays, that is the possibility to create a representation of the university in the municipality, in order that professors and students, can have the possibility to be together with the municipality administration, as also with the legislative and judicative powers.

4. Conclusion

The Team of the *Escola Politécnica* of UFRJ was focusing on the contribution of a solution for the metropolises urban situation. It is intended to establish and improve the specific and substantial improvement of the urban development of the smaller cities, which are responsible for the immigration towards the Metropolitan Area.
With the participation of the University of Applied Sciences of Berlin (Technische Fachhochschule Berlin) the *Escola Politécnica* of UFRJ is supporting the municipality of São José do Vale do Rio Preto with information on how to develop and establish a sustainable urban planning structure.
The work with the community is becoming stronger with the support of a project that has been run by the European Community for Latin America - ALFA Project, called AURORA ("Architectural and Urban Research of Regional Agglomerations" - www.alfa-aurora.net)[6]. Results of this project can be seen in Michalka (2007a) and Ten Hold et al. (2007).
The exchange of ideas about the reality of the municipality and the intention to elaborate a sustainable development with the contribution of international experts is a very important step towards the future for the community. This is also possible with the partner universities to gain a better understanding about the "every day problems" of a small community in Brazil. Handling this process and giving contributions that can be used by the administration in order to take more appropriate decisions on planning, land use and infrastructure, is what the universities are helping to develop.

Nowadays there are six priority action lines:

- master plan;
- land use;
- technical infrastructure;
- social infrastructure;
- environmental planning;
- local sustainable development;
- regional development.

It is very important to clarify the objectives of the team work of university and community:

- the university provides the community with the necessary information and consults in order to enable the community to make wise decision within the projects for a sustainable development.
- it is the responsibility of the community to make the decisions; because the community knows better about its own necessities.

The aim is to develop sustainable solutions for the smaller cities, as an answer to the huge pressure by the immigration towards the Metropolises, in order to integrate the city in the county's necessities in order to reduce the migration from the small cities and the rural areas.

The smaller cities have better opportunity to implement a sustainable development and growth of their municipality. The scale of these cities can still be influenced and have several chances to establish a planning structure. This focus may contribute to solve the tasks of an accelerating agglomeration of inhabitants that produces tremendous areas of informal settlements like the ones in Rio de Janeiro's metropolitan area.
The objective of the *Escola Politécnica* of UFRJ is to work with this new point of view. This means to handle the problems of the Metropolis by going to one of the most important origins of the problem, which is the migration towards the Metropolises.

5. References

Capra, F. (1997). *The Web of Life*, Achor
da Silva, J. X. (2004). *Geoprocessamento para Análise Ambiental*, Editora Bertrand
McHarg, I. L. (1992). *Design with Nature*, John Wiley & Sons
Michalka Jr., C. (2007a). Sustainable development: Municipality/Public University partnership. In: Marszal, W. Z. (Org.) *Metropolises and Metropolitan Areas - Structures, Functions and Role (based on Polish and Brazilian Experience)*, Warsaw: Polish Academy of Sciences - Committee for Spatial Economy and Regional Planning

Michalka Jr., C. (2007b). *Concepts for Revitalization an Development of the São José do Vale do Rio Preto Municipality, Rio de Janeiro, Brazil.* Research and Workshop Results of the ALFA-AURORA Network, ALFA Program of the European Commission Project Number AML/B7-311/97/0666/II-0369-FCD-FI-FC

Safatke, A. & Bernardi, B. (2008). Territórios Ocupados, In: *Informação para o Novo século,* Fundação Getúlio Vargas, Centro de Estudos em Sustentabilidade da Escola de Administração de Empresas de São Paulo, N. 16

Ten Hold, F.; Ledo, I. & Trautwein, M. (2007). *Low Cost Housing on Slopes. Evaluation of a HVA research in Brazil 2006/2007.* Hogeschool van Amsterdan

9

Computational Tools applied to Urban Engineering

Armando Carlos de Pina Filho, Fernando Rodrigues Lima,
Renato Dias Calado do Amaral
Federal University of Rio de Janeiro (UFRJ)
armando@poli.ufrj.br, frlima@poli.ufrj.br, natodias@poli.ufrj.br
Brazil

1. Introduction

The objective of this chapter is to present some of the main computational tools applied to urban engineering, used in diverse tasks, such as: conception, simulation, analysis, monitoring and management of data.

In relation to the architectural and structural project, computational tools of CAD/CAE are frequently used. One of the most known and first software created to Personal Computers (PCs), with this purpose, was the AutoCAD by Autodesk. At first, the program offered 2D tools for design assisted by computer, presenting technical and normalisation resources. After that, the program started to offer 3D tools, becoming possible the conception and design of more detailed environments. The program is currently used for construction of virtual environments (or virtual scale models), being used together with other programs for simulation of movement and action inside of these environments.

Another software very used currently is the ArcGIS, created to perform the geoprocessing, in which tools and processes are used to generate derived datasets. Geographic information systems (GIS) include a great set of tools to process geographic information. This collection of tools is used to operate information, such as: datasets, attribute fields, and cartographic elements for printed maps. Geoprocessing is used in all phases of a GIS for data automation, compilation, and management, analysis and modelling of advanced cartography.

In addition to the programs of CAD and GIS, other interesting technology is related to Building Information Modelling (BIM), which represents the process of generating and managing building data during its life cycle using three-dimensional, real-time, dynamic building modelling software to decrease wasted time and resources in building design and construction. Some of the main software used for BIM are Autodesk Revit Architecture and Vico Constructor.

Computational tools for monitoring and management are very important for the urban development. Several urban systems, such as: transports, water and sewerage system, telecommunications and electric system, make use of these tools, controlling the processes related to each activity, as well as urban problems, as the pollution.

Therefore, in this chapter we will present details about these technologies, its programs and applications, what it will serve as introduction for the other works to take part in this book, many of which use such computational tools for study and solution of urban problems.

2. CAD (Computer-Aided Design)

It is a technology largely used in the conception of projects of Engineering and Architecture. It consists of a software directed to the technical drawing, with several computational tools. Amongst the areas in which the CAD is applied, we have the Urban Engineering.
Urban Engineering studies the problems of urban environments, emphasising the creation of planned environments to be sustainable, considering the balance of economic, territorial, and social factors. The infrastructure urban systems are subject of study, searching to optimise the planning of the environment, sanitation sectors, transports, urbanism, etc. In this context, we can observed the use of CAD programs to assist urban projects.
In respect of development of CAD software, we observe that without the postulates of the Euclidean Mathematics (350 B.C.) it would not be possible to create this computational tool.
Thousand of years later, more specifically at the beginning of the 60th decade of the 20th century, Ivan Sutherland developed, as thesis of PhD in the Massachusetts Institute of Technology (MIT), an innovative system of graphical edition called "Sketchpad". In this system, the interaction of the user with the computer was perform by "Light pen", a kind of pen that was used directly in the screen to carry through the drawing, together with a box of command buttons. It was possible to create and to edit 2D objects. Such system was a landmark in computer science and graphical modelling, considered the first CAD software.
In the beginning, the use of CAD software was restricted to companies of the aerospace sector and automobile assembly plants, as General Motors, due to the high cost of the computers demanded for the systems. Such software were not freely commercialised in the market. The Laboratory of Mathematics of MIT, currently called Department of Computer Science, was responsible for the main research and development of CAD software. In other places, as Europe, this type of activity was started. Other prominence developers were: Lockheed, with CADAM system, and McDonnell-Douglas, with CADD system.
From the 70th decade, CAD software had passed to be freely commercialised. The first 3D CAD software, CATIA - Computer Aided Three Dimensional Interactive Application, was developed in 1977 by French company Avions Marcel Dassault, that bought the Lockheed, revolutionising the market. The investments, as well as the profits, vertiginously grown. In the end of the decade, programs for solid modelling already existed, as, for example, the SynthaVision of the Mathematics Application Group, Inc. (MAGI).
From 1980, with the development of the first Personal Computer (PC), by IBM, the Autodesk released, in November 1982, the first program of CAD for PCs, the "AutoCAD Release 1". In 1985, the Avions Marcel Dassault released the second version of CATIA. In this same decade, the workstations (microcomputers of great efficiency and high cost, destined to technical applications) were developed, using the operational system UNIX.
In the 90th decade, specifically in 1995, the SolidWorks company released the SolidWorks 95 3D CAD, revolutionising the market for used the operational system Windows NT, while the majority of the programs developed was destined to UNIX. In consequence of this, SolidWorks 95 demonstrated to be a software with good relation of cost-benefit, when compared with the competitors, excessively expensive.

In the following years to present time, the technology comes being improved and the software became very accessible around the world, with open access versions (freeware).
An important application of the 3D CAD programs is the creation of virtual environment, also known as electronic or virtual scale models (Fig. 1). Such application is largely used in architecture projects.

Fig. 1. Example of virtual scale model: *Hospital Metropolitano Norte*, Pernambuco, Brazil (http://acertodecontas.blog.br)

2.1 Working with CAD

As previously said, we had a great development of CAD software in the last decades. Amongst the main programs of CAD, the AutoCAD (http://www.autodesk.com.br) is distinguished. The software developed by Autodesk had its first version released in 1982, and recently, the Autodesk released the AutoCAD 2010.
The AutoCAD (Fig. 2) is a 2D and 3D modelling program with several applications, such as: mechanical, civil, electric, and urban engineering projects; architecture; industrial manufacture; and HVAC (heating, ventilation and air conditioning). It is important to notice that the AutoCAD is also largely used as tool in academic disciplines of technical drawing.

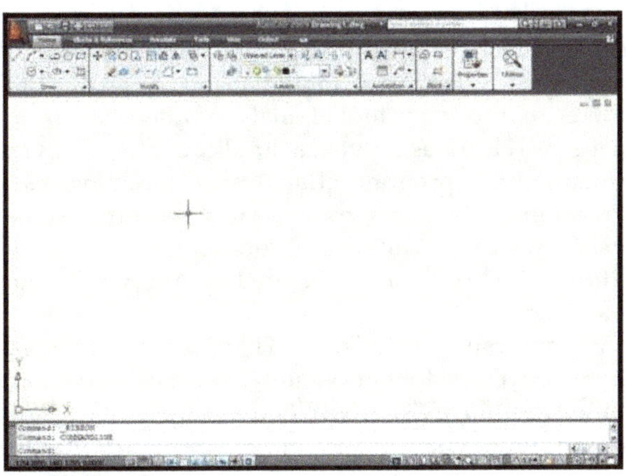

Fig. 2. Interface of AutoCAD software

AutoCAD have commands inserted by keyboard, making possible a practical creation of entities (elements of the drawing), at the moment of the conception of the desired model, optimising the work of the designer. Such commands substitute the necessity of navigation with the mouse to manipulate the toolbars.

The program generates diverse types of archive, which can be exported to other programs. Some examples: DWG (*.dwg); 3D DWF (*.dwf); Metafile (*wmf); Encapsulated (*.eps); and Bitmap (*.bmp). DWG archive is an extension shared for several CAD programs. AutoCAD is capable to import archives of the type 3D Studio (*.3ds), from Autodesk 3D Studio Max.

User of AutoCAD is able to associate with your projects, programs made by programming languages, such as: Visual Basic for Applications (VBA), Visual LISP e ObjectARX.

Another CAD software largely known is the SolidWorks (http://www.solidworks.com). Developed by SolidWorks company, from group Dassault Systèmes, is a 3D CAD program for solid modelling, generally used in the project of mechanical sets (Fig. 3).

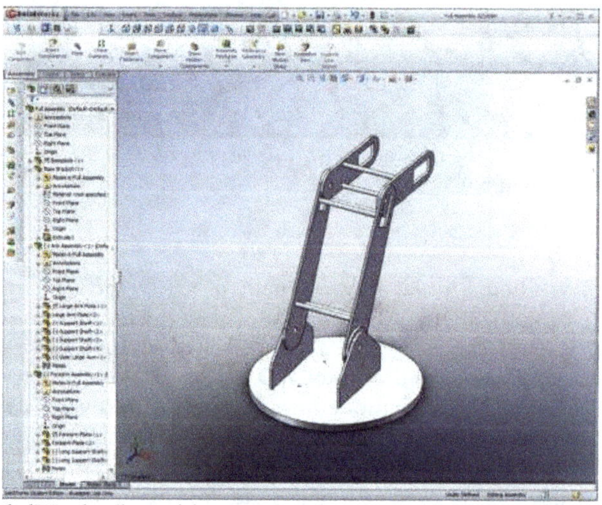

Fig. 3. Project in SolidWorks (http://www.danshope.com)

SolidWorks can also be used as CAE software (Computer-Aided Engineering), with simulation programs, such as: SolidWorks Simulation, and SolidWorks Flow Simulation.

SolidWorks Simulation is an important tool of analysis of tensions in projects. The program uses finite element methods (FEM), using virtual application of forces on the part.

SolidWorks Flow Simulation is a program of analysis of draining, based on the numerical method of the finite volumes. This program allows the professional to get reasonable performance in analysis of the project under real conditions.

SolidWorks is compatible with DWG files generated by AutoCAD, being able to modify 2D data or to convert into 3D data.

Other interesting CAD programs include: CATIA (Computer-Aided Three-dimensional Interactive Application), developed by Dassault Systèmes and commercialised by IBM (http://www.3ds.com), and Pro/ENGINEER, developed by Parametric Technology Corporation (http://www.ptc.com).

2.2 Application of CAD

CAD software have as main use the aid in projects of Civil Engineering and Architecture for urban environment, such as: buildings, roads, bridges, etc (Fig. 4).
CAD also is widely used in the project of transmission lines of electric energy. Such practice consists in optimise the allocation of transmission towers and wires, in accordance with the technical norms. An important characteristic is the topography of the land.

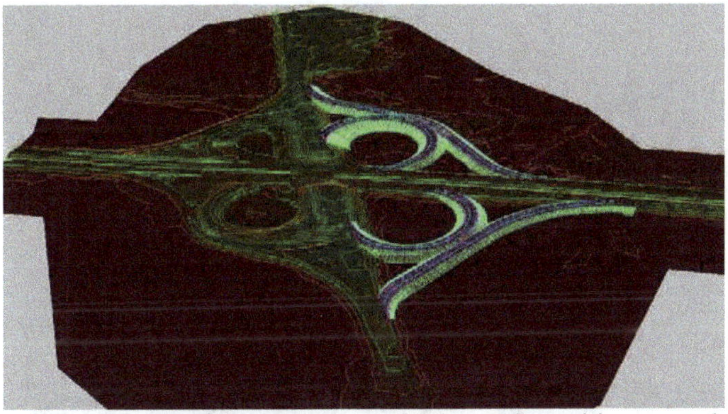

Fig. 4. Example of project of Civil Engineering - a highway (http://usa.autodesk.com)

Other applications in Urban Engineering include: the maintenance and update of sanitary networks, and the environmental recovery in urban areas. In the first case, CAD is used to update the database of the sewer network of the city, supplying detailed information. In the second case, CAD is used for mapping of a region, with the aid of a GPS system (Global Positioning System), identifying environmental delimitation (sources of rivers, roads, buildings, etc)(Mondardo et al., 2009).
There are several other applications of CAD in urban systems and areas related to Urban Engineering, and it is important to note, in practical terms, that CAD is nearly always associate to other technology: GIS (Geographic Information System), that it will be seen to follow.

3. GIS (Geographic Information System)

Engineering problems were on the last 40 years gradually directed to employ computerised solving techniques. Precision and increasing speed for calculating multi-variable operations are a good reason to use computational resources, but the quite unlimited possibilities to organize, simulate and compare data turned computer sciences on a strong allied for research and design activities.
The final claim to say that now we are living in an information systems age is the large accessibility of hardware and software, the diffusion of personal systems and all related facilities: servers, networks, telecommunications, etc.
An information system can be defined as an organised quiver of tools and data that can be used to answer on a systematic way questions structured by specialists. As these questions can be classified in patterns, it should be possible to build on artificial intelligence to make the system learn and deliberate by itself.

If the answer to a problem employs variables associated to geographic information, it's recommended the use a dataset structure to implement and model graphic objects that represents all on earth, natural or artificial. A Geographic Information System (GIS) is a set of tools that work with data presenting three basic concepts (Fig. 5): Geodatabase, Geovisualization and Geoprocessing (Harlow, 2005).

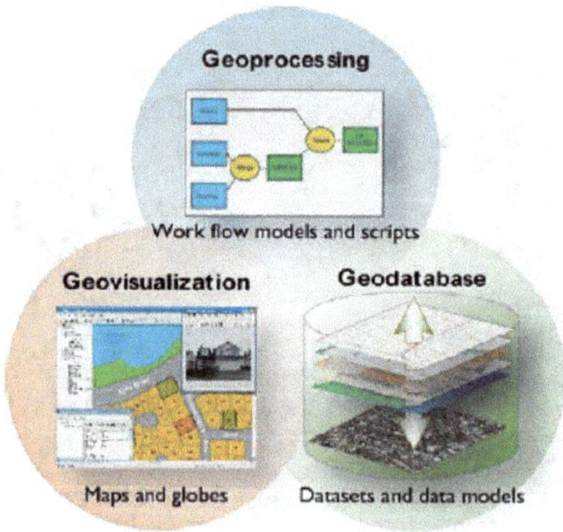

Fig. 5. Schema extracted from ArcGIS Reference manual showing the three views of GIS

Geodatabase represents the set of spatial data that can be expressed by rasters, vector features, networks, etc., and every rule to control their creation and management. Geovisualization is an action performed on spatial data by intelligent maps and views, from which we can view the database for querying, analysing and editing. Geoprocessing is the term used to designate operations on datasets that obtain outputs of analyses and generate new information.

Some engineering projects have territorial themes as industrial projects, social benefits, general infrastructure, logistics, demography, and other geo/urban/environmental aspects. On those cases the solution involves studying geographic elements and their available information, in order to perform technical analyses. So an information system for geographic data organisation, visualisation and processing will be appropriate to those problem solving. To be efficient as a GIS, the system must perform some general tasks: acquire, convert, organise and project the geographic elements; import, organise and extract imagery, numeric and textual information; process geographic elements and information with data integrity and operational efficiency; display appropriately the data and related operations (geoprocess techniques); perform simulations and comparison of alternatives; present support for program language and custom computational routines; generate new data based on selected results; publish maps and all sort of documents for project discussion; and permit data interchange with other systems.

As we can observe, GIS is designed to manage spatial data, and the geographic representation of this data can use many types of elements for plotting the information (Fig.

6): vector based features classes, as points, lines and polygons; raster datasets, as digital elevation models and imagery; networks, as roadways, pipelines, hydrology and other interlaced elements; survey measurements, as topographic annotation; and other kind of information, as postal codes, address, geographic place names, etc.

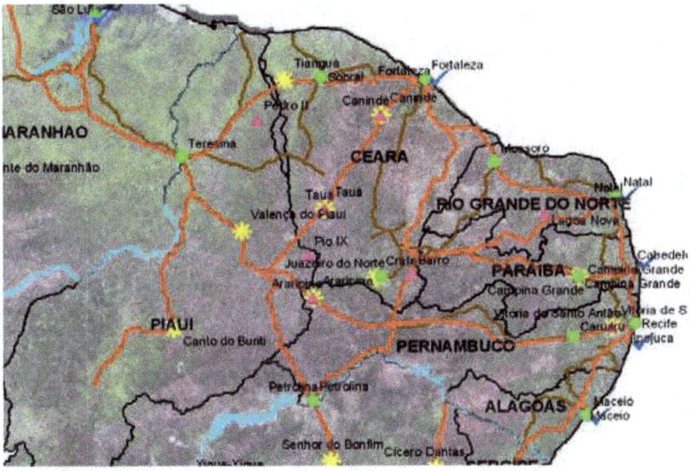

Fig. 6. Vector features overlays raster satellite image

These elements can be organised by layers, and could be selected by pointing or grouping for edition tasks or custom display. The selection methods could also be performed from spatial analyses or statistic classification. Georeferenced co-ordinates and related data tables of GIS elements help to improve these tasks.

Geographic data representation has integrity rules (Harlow, 2005), performed by spatial relationship patterns between elements, as topologies and networks. Topologies are used to manage boundaries behaviour, to apply data integrity rules, to define adjacency and connectivity properties, to structure creation and edition of new geometry, and to express other topological operations. They are used to represent area contours, parcels, administrative boundaries, etc. Networks are used to represent graphs and their connections, controlling paths, barriers and flows. They are used to represent behaviour of pipeline, transportation, traffic, etc.

Although organisation and management of spatial data can be well attempted with modern GIS programs, there is until an important aspect: how to deal with data quality. The cartographic databases can be generated from old charts or maps digitalisation, or from satellite and aerial imagery treatment. The numeric and textual databases must be converted into tables, and quite often comes from census and researches output. A great variability of data procedures can be observed world-wide when integrating data obtained from different fonts, places and scales. The periodicity of data actualisation is another deal to GIS users.

The problems don't result ever from confidence, trusted fonts may have different methodological approaches, and personal interpretation can also give different valid outputs. Professional development of GIS operators can help them to detect, evaluate and work that variability, and a methodological approach is needed to treat it suitable to each research task.

3.1 Working with GIS

Many users can be satisfied on using GIS as a dataset management tool for generating maps and classify data, but nowadays GIS is turning on a knowledge approach, where models incorporate advanced behaviour and integrity rules. The ultimate development on GIS procedures is directed to intelligent use of geoprocessing for built, explore and share the possibilities of geographic information. Users now are able to structure schemas and workflow models in order to improve their geoprocessing tasks, as import, check, integrate and compose data (Fig. 7).

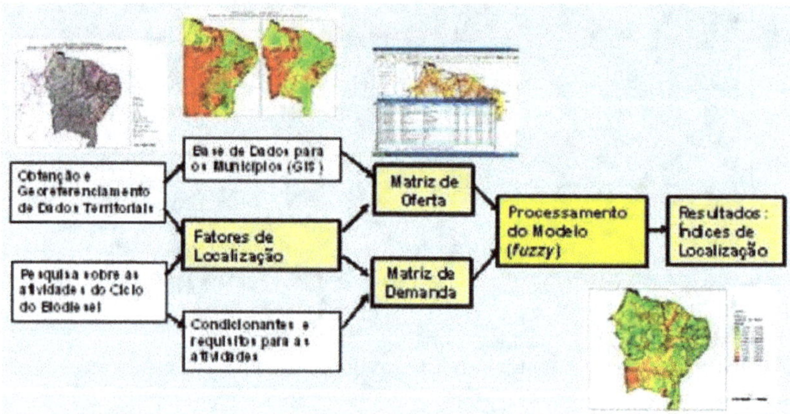

Fig. 7. Example of a workflow model for GIS based research on industrial location

As GIS is the best way to work data from local to global level, an efficient DBMS (Data Base Management System) is needed to perform data integration, actualisation, access and sharing. As result, GIS catalogue portals based on Web nodes are increasing in number and their interoperability is part of a concept called SDI (Spatial Data Infrastructure). Servers are used to host enterprise GIS and their databases, and to provide multi-user access. Geographic Databases are employed to control and develop published data, as maps, features and tables. They are known as Geodatabase, have a proper logic to work with datasets by applications and tools, and perform access and management tasks.

But GIS capabilities can also provide single users to customise their data. A Personal Geodatabase (PGDB) is an example of option to collect and organise features and tables attempting to user needs, using desktop computers at low cost and with feasible results. If you are a adviser or researcher and are in charge of studying urban problems, you can go ahead on mounting your PGDB, however some steps must be observed.

The start point is to structure correctly your problem, identifying the factors and conditions that impacts on, a methodological approach to face it, and a technical procedure to get alternatives and produce results.

First, you must study what kind of information you need, identify the sources and think about layer and features organisation. Next, you must acquire geographic data from GIS portals or institutional sources. Many research and administrative institutions provide download of vector and raster data from their DBMS, or send it by request. If there is no available geographic data, it will be necessary to digitalis existing map and imagery, but for this task is recommended a professional with advanced knowledge of geodesics, cartography and geoprocessing.

After getting the appropriate geographic information is important to know that vector data is usually related to a table, which has a column whose contents link the graphic representation to a register. Raster image has pixel position attached to a co-ordinate value. Vector features as point, line or polygon has as code number for the system link requirements, but can also have a code for geographic cadastral purposes (Fig. 8). Geocode is a tendency on GIS procedures and has the advantage to make easy later joins and relates of table data with none geographic plot.

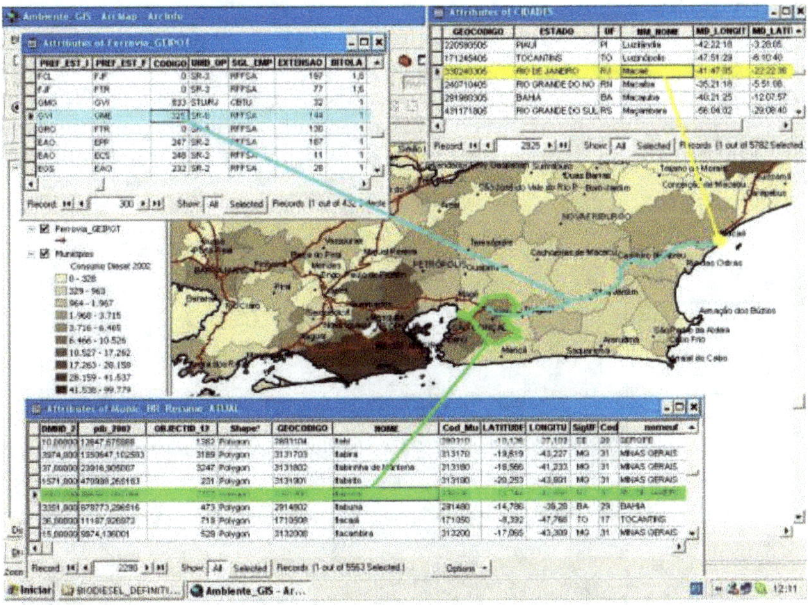

Fig. 8. Vector features as point, line and polygon with associated table containing geocode

In other words: if you get a basic data of shapes with related table presenting geocode column you can aggregate new data from other ordinary tables that has also this geocode column. GIS also enables visualisation of each element by selecting it from geocode, and permits editing the tables to insert new columns containing yours own information.

Second, you must organise your features and tables in a dataset, defining co-ordinate systems and importing independent features and tables to the PGDB. This modality of data organisation provides more security and flexibility, increasing edition and analysis tasks. Working with stand alone features can face restrictions that are not present on a PGDB structure, as it works more properly with layers, overlays, projections and co-ordinates.

Third, you must know what to do to improve your queries on GIS ambience. It is a lost of potential to use a GIS only for data visualisation or map creation, there is more than this. Both DBMS or PGDB can generate data performing spatial analysis or statistic classification. As you have the demands of your research well structured, GIS can help you to answer by crossing multi-layer information, selecting and editing data from SQL (Structured Query

Language) statements and processing new features containing partial and conclusive results.

Finally, you must obtain a valid output for your problem solving, and communicate it to others on a suitable way. GIS can help you on producing thematic maps, analytical graphs and technical reports. You can also get community, representatives and specialists to work in a participative mode using GIS to generate and validate output of decision sessions. Some people have difficulties to identify and interpret geographic elements, and GIS can highlight and detach text and visual information for making it easier.

3.2 Application of GIS

GIS technology is much used in Urban Engineering to analyse, in a detailed way, characteristics related to urban planning. In addition to CAD, GIS presents solutions for several problems, and it is applied, in a integrated way, in projects of Civil Engineering and Architecture, including the most diverse urban systems (Fig. 9), making possible the maintenance and update of service networks, as well as the environmental recovery in urban areas.

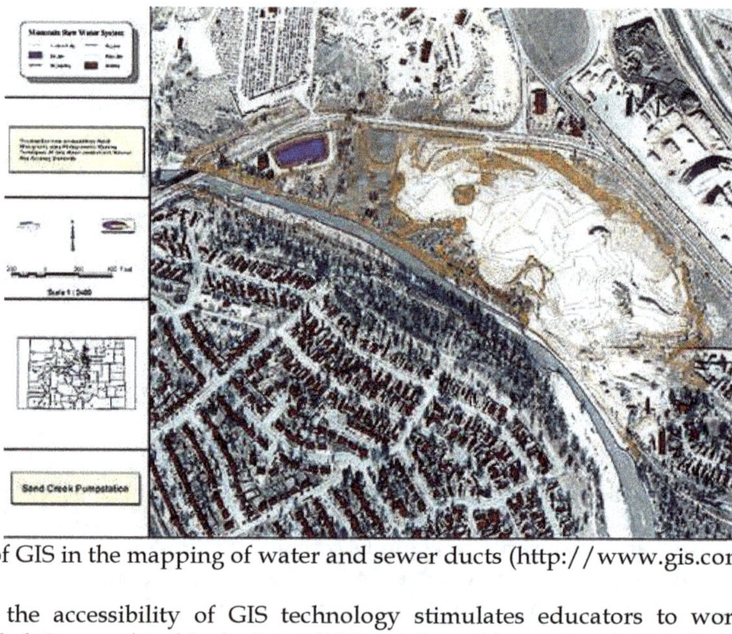

Fig. 9. Use of GIS in the mapping of water and sewer ducts (http://www.gis.com)

Nowadays, the accessibility of GIS technology stimulates educators to work in a new concept, called Geographical Inclusion, which can be performed on basic education class in order to provide young students with geographic visualisation and interpretative capabilities. We are living a age of saving resources, environmental care and sustainable actions, and GIS with his solving problem design and participative net work potential is the most strong partner in managing data for this purposes.

Concluding the technologies presented in this chapter, we will see to follow the BIM technology (Building Information Modelling), that it represents, in a certain way, an evolution of CAD technology, previously presented.

4. BIM (Building Information Modelling)

It is a technology that consists in the integration of all types of information related to conception and execution of a project of Civil Engineering. Such information, stored in efficient database, not say respect only to design or to modelling of plants and virtual environments, but also to management of execution time of project, geographic information, quantification of material used in all building, detailing of the constructive processes, sustainability, etc. In short, the technology makes possible that the work team has an integrated vision of the project. This allows, for example, that engineers and architects idealise and execute the project sharing the same base of information. This technology has been spread together with the practice of Urban Engineering.
In a certain way, BIM is seen as an evolution of 3D CAD techniques. In fact, this technology is defined as 4D CAD, where the fourth dimension is not physical, but the set of information that go beyond the engineering concepts, used in the development of the project.
The use of BIM can mean an effective optimisation of time and increase of the productivity levels. Other important characteristic is the easiness to perform modifications in the project, in any phase of execution. BIM makes possible the meeting of information, such as: the documentation of licensing for building, the established environmental conditions, and other legal aspects that are of extreme importance for execution of the project. Thus, the technology allows to greater efficiency in the taking of decisions during the elaboration of the project, easiness in the emission of building documents, establishment of deadlines, estimate of costs, information about the analysis of risks and management of the operational conditions of the installations.
Using a CAD software in an engineering project, the designer inserts detailed specifications through the headings, for example: specification of the material used in the confection of a wall, manufacturer of the material, necessary amount. In the case of BIM technology, such information is directly inserted in the drawing at the moment of the modelling.

4.1 Working with BIM

In BIM technology, a set of tools provided by one or more software is used for: modelling of surfaces; modelling and structural calculation; management of the building; manufacture management; environmental analysis; estimate of costs; and specification.
Autodesk Revit Architecture (usa.autodesk.com)(Fig. 10), is one of the main BIM software, having: tools of 2D and 3D drawing; co-ordinated database, in such way that alterations performed in the information are automatically update in all model, reducing the possibility of errors and/or omissions; associative sections of divisions table; libraries of details, that can be created and be adapted to the patterns of the project team; parametric components, that function as an open graphical system for design concern and shape creation; inventory of materials, that allows the calculation of detailed amounts of material, updating while the project evolves, on the basis of parametric alterations; etc.
There are other BIM programs by Autodesk, as Autodesk Navisworks (usa.autodesk.com), that it does not present tools of environment modelling, being destined to the revision of 3D projects or visualisation of models, that is the case of the freeware NavisWorks Freedom. The main tools include: aggregation of files and 3D data; revision tools; creation of 4D table; object animation; management of interference and detention/correction of conflicts in the project; exportation of DWF files (used in CAD programs); and navigation in real time.

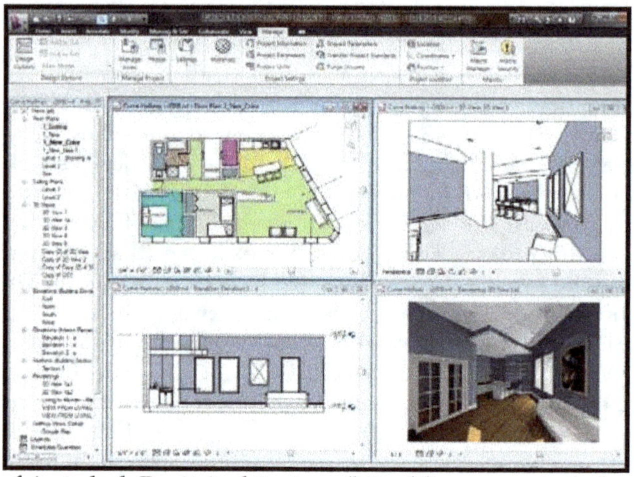

Fig. 10. Interface of Autodesk Revit Architecture (http://images.autodesk.com)

Other popular freeware is the Google SketchUp (http://sketchup.google.com)(Fig. 11), much used in the academic area, presenting modelling by means of surfaces. Such software presents limitations compared to the programs already cited. SketchUp works efficiently with information related to the localisation, size and design, reason for which is used in the confection of models that can be exported to programs, as for example, the Autodesk NavisWorks Review.

Fig. 11. Interface of Google SketchUp (http://www.crackvalley.com)

There are several other programs related to BIM technology, as for example, Vico Constructor (http://www.vicosoftware.com), presenting diverse characteristics and resources, such as: the structural analysis of the building; the constant update of the information, correcting possible errors of execution; the estimate of costs of the enterprise; etc.

4.2 Application of BIM

As well as CAD and GIS technologies, BIM presents a series of applications in the area of Urban Engineering, and currently it comes substituting CAD, in a effective way, because it presents advantages in relation to the management of the projects.

A current example of BIM application is the National Centre of Swimming of Pequim, China (en.beijing2008.cn/46/39/WaterCube.shtml). Seat of the competitions of swimming during the Olympic Games of 2008, known as Water Cube, the place have a useful area of 90,000 m^2, five Olympic swimming pools and capacity for 17,000 spectators, and BIM was used in all phases of the project (Fig. 12 and 13).

Fig. 12 and 13. Assembly of the structure of the Water Cube, and aerial photo of the place (http://comunicacaoexponencial.com.br)

Other example of BIM application is the International Airport Maynard Holbrook Jackson Jr., Atlanta, United States (Fig. 14). This airport is in construction phase with a stipulated deadline for 2011. In this project, of great magnitude, BIM is extremely necessary in the optimisation of execution time, since the old airport of Atlanta is overloaded. The estimated cost of the enterprise is approximately US$ 1.4 billion (Ford, 2009).

Fig. 14. Model of the Airport (http://bim.arch.gatech.edu)

5. Conclusion

This chapter looked for to present the main details on three technologies much used in Urban Engineering: CAD (Computer-Aided Design); GIS (Geographic Information System); and BIM (Building Information Modelling). As it can be seen, each one of them presents specific characteristics and with diverse applications in urban projects, providing better results in relation to the planning, management and maintenance of the systems.

In relation to presented software, it is important to note that the authors do not have any connections with the cited companies. The programs were shown only as computational tools that use the presented technologies, and there are many other commercial software and freeware that can be used in works involving CAD, GIS or BIM. Therefore, the work presented here does not represent any intention of marketing for no one of cited software and/or companies.

6. References

Autodesk (2006). Orange County Sanitation District - Customer Success Story. *Autodesk Infrastructure Solutions*

Autodesk (2009). Langan Engineering & Environmental Services - Customer Success Story. *Autodesk Infrastructure Solutions*

Ford, K. (2009). Maynard Holbrook Jackson Jr. International Terminal. *Holder Construction Group LLC*, Georgia Tech

Harlow, M. (2005). *ArcGIS Reference Documentation*. ESRI: Environmental Systems Research Institute Inc., Redlands

Kymmell, W. (2008). *Building Information Modelling - Planning and Managing Construction Projects with 4D CAD and Simulations*. The McGraw-Hill Companies, Inc

Mondardo, D., Bellon, P. P., Santos, L. B., Meinerz, C. C. & Haoui, A. F. (2009). Proposta de Recuperação Ambiental na Área Urbana da Microbacia do Rio Ouro Monte. *2nd International Workshop - Advances in Cleaner Production*, São Paulo, Brazil

Sutherland, I. E. (2003). *Sketchpad: A man-machine graphical comunication system*. Technical Report 574. University of Cambridge, Computer Laboratory, p. 20

10

Sound Pressure Measurements in Urban Areas

Fernando A. N. Castro Pinto
Federal University of Rio de Janeiro (UFRJ)
fcpinto@ufrj.br
Brazil

1. Introduction

The assessment of noise pollution is of prime concern in modern urban planning. Nevertheless, even in the case of the use of numerical simulations it must rely on actual measurements. In order to make reliable noise evaluations one must take into account the position of the measurement but also have a thorough knowledge of the equipment used like microphones, pre-amplifiers, and analysers. Their correct configuration is mandatory for a correct interpretation of the obtained results. The signal processing techniques ranging from detector types to the quantification of equivalent levels is discussed in this chapter.
A correct analysis of a sound event, the background noise from the vehicle traffic on streets and avenues, noise coming from industrial facilities or from air conditioning equipment, involves various parameters as sketched in Fig. 1. Not all of them can be evaluated and quantified with standard measurement equipment and techniques. Nevertheless it is possible do establish metrics for comparison of different situations and for the determination of noise thresholds in laws and regulations to protect the population from too high levels.

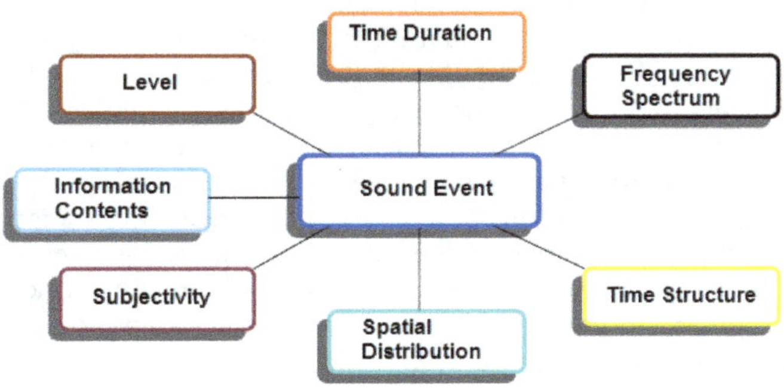

Fig. 1. Sound event parameters

2. Sound Pressure Level

The assessment of the noise pollution in urban areas involves the quantification of the sound exposure of the population. This task mat seem at a first glance relatively simple with the use of sound level meters, equipment which is readily available in different types. Nevertheless, in order to correctly evaluate the measurements, a thorough understanding of the acoustic phenomena involved and of the instrumentation itself is needed.

Besides the data acquisition equipment the signal processing associated with the measurements, i.e. the configuration of the sound level meter, will lead to the measurement of different parameters that are used as indicators of the noise impact.

The basic problem faced here is related to the fact that sound is a sensation caused by the pressure fluctuation occurring in our hearing system. The outer ear performs the task of gathering the pressure waves leading them to the middle ear, the middle ear in turn transforms the pressure fluctuations on the tympanum membrane into waves inside the inner ear, the cochlea, where a frequency analysis is done and the signals are then sent to a further processing in the brain.

As can be seen, a major problem arises from the engineering feasibility of measuring pressure fluctuations to actually quantify a *sensation* produced inside the brain. It is not difficult to devise the measurement of the pressure fluctuations. The microphones being a membrane which is excited by the ambient pressure transforming it into a fluctuating force over a sensing element. Ideally this sensor shall respond equally well to different kinds of fluctuations., i.e. to different frequencies of excitation.

Although desirable from the point of view of a good sensor this flat response is quite different from the actual response of our hearing system. Therefore it is necessary to adapt the results of the microphone measurements in order to get values that can be correlated to our sound sensation.

Another problem is that this sound sensation is not so simple, but is full of different aspects such as loudness, pitch, tonality, roughness, fluctuations, etc., besides being influenced by the duration of the noise event. Each of these aspects may lead to a different perception of the sound and to different impacts of the noise exposure. The definition and quantification of these different aspects of the sound sensation are studied by the field of psycho-acoustics, which is beyond the scope of this text. For the assessment of the noise pollution we shall concentrate on the sensation of loudness or volume which is mainly associated to the amplitude of the pressure fluctuations of the sound waves.

In general the human body reacts to actual *physical stimuli*, such as pressure fluctuations, light, temperature, etc., to create a sensation in a relative way. It means that the increase in the sensation is related to the relative increase in the physical stimulus. Weber and Fechner formulated this empirical law as a logarithmic function of the stimulus intensity (Schick, 2004). This intensity is in turn represented by the energy content of the stimulus. In the case of sound this energy is represented by the square of the pressure fluctuation.

One step towards the quantification of the sensation from pressure measurements is the evaluation of this energy through the definition of a mean pressure which, over a suitable time interval *T*, would yield the same acoustic energy of the event of interest. This is done through the RMS (Root Mean Square) value $\overline{P}_{RMS}$ of the time varying pressure fluctuations $p(\tau)$, as in equation (1).

$$\overline{P}_{RMS} = \sqrt{\frac{1}{T}\int_{\tau-T}^{\tau} p(\tau)^2 \, d\tau} \tag{1}$$

Further, one must build the equivalent of the logarithmic function for sensations, going from an actual pressure measurement as above to a dimensionless quantification of the so called *Sound Pressure Level* or *SPL*, presented in equation (2).

$$SPL = 10\log\left(\frac{\overline{P}_{RMS}}{P_{ref}}\right)^2 \tag{2}$$

where P_{ref} is a reference pressure value. The calculated SPL value is expressed in decibels, abbreviated dB, and the commonly, standardised, used value for P_{ref} is 20 μPa. The choice of a suitable time interval T for the measurement, will be addressed further in this chapter.

Another step in order to approximate the sound sensation from the sound pressure measurements is related to the frequency response of the ears. We are neither able to hear fluctuations with high frequencies, above about 20kHz, nor low frequencies, below about 30Hz. Of course the actual values are individual and may be influenced by age, diseases, and long exposure to high noise levels, among other factors. This can be translated into a sensitivity which varies with frequency. Actually the frequency dependency is also related to the loudness of the sound itself and one may construct curves representing equal sensitivity, the so called isophonic curves. Based on the pressure fluctuations of different frequencies and amplitudes that cause the same hearing sensation to a single person they are constructed from a statistical analysis of many of these measurements.

In contrast the ideal measurement microphone would show equal sensitivity to all frequencies, independently of the sound pressure level being assessed. This final step to evaluate noise consists in the *distortion* of the measured pressure signal through a filter, analogic or digital, that resembles the inverse of the sensitivity of the ears. Four such weighting filters are defined in international standards (IEC, 2002) and designated by the letters A, B, C and D. The most important for noise pollution assessment are the A and C filters, which approximate the sensitivity at levels near 40dB and 90dB respectively. Figure 2 shows the frequency characteristics of the different weighting filters.

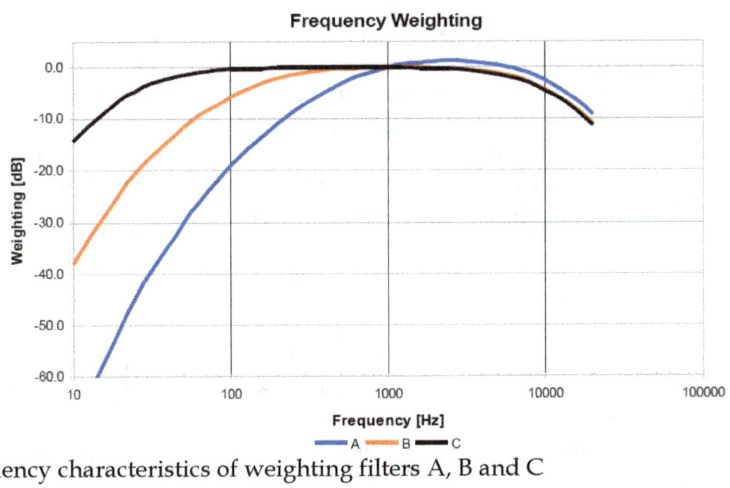

Fig. 2. Frequency characteristics of weighting filters A, B and C

Weighting curve D is mainly used to assess airport and aircraft related noise. Curve B, although interesting, since it resembles the inverse isophonic of about 70dB, a common encountered level, is seldom used. Curve C is preferred to assess impact noise and high levels. It is possible to recognise the low sensitivity for low frequencies, specially below 30Hz, and high frequencies, above 18kHz.
Nevertheless, due to the non-linear nature of the RMS calculation the filtering process, the frequency weighting, must actually be done previously, i.e. before the RMS calculation takes place. Firstly the sensitivity of the microphone is corrected to reflect the human sensitivity through the frequency weighting, after that the RMS-value is calculated and finally the level is found through the logarithmic function.
The finally obtained sound pressure level involving all described steps is expressed in dB(A), dB(B), dB(C) or dB(D) according to the filter used. In the case of no filtering it is advisable to express the measurements as dB(L), from linear or no filtering, or dB(F), from a flat response frequency filter, stating clearly that no frequency weighting was used.
It is clear from the above that the specification of a Sound Pressure Level is a mere trial to approximate the sound sensation.
Since the frequency weighting depends on the frequency spectra of the sounds being measured, without knowledge of these spectral contents of the measurements, it is not possible to convert a value expressed in dB regardless of the suffix, A, B, C, D or L (or F), into another one with a simple additive or multiplicative factor. Therefore it is sometimes advisable to measure and record the sound spectra as well.
Yet there are still deeper aspects related to the perception of loudness such as frequency and temporal masking which are not addressed by the described procedure. This can be improved through the use of other volume metrics such as the *Loudness*, defined in the ISO 532 standard (ISO, 1975), nevertheless it is still not practice in noise pollution assessment, neither is the evaluation of most other aspects of sound perception. Only the tonality is taken into account and will be discussed further in the spectral analysis of the sound, since it is related its frequency distribution.

3. Instrumentation

The instrumentation used to evaluate the SPL as described earlier involves the sensor, microphone, to capture the pressure fluctuations, a preamplifier to adequate the, low power, electrical signals from the microphone to a meter or a frequency analyser to perform the RMS calculation, the frequency weighting and analysis, showing the final obtained values. Often it is desirable to store the measurements, along with a time stamp, and/or to transfer them to a computer either locally or remotely. The accuracy of the whole measurement chain, and the adjustment of the microphone sensitivity, is achieved with the help of an acoustic calibrator.

3.1 Microphones and Preamplifiers

The microphones used in noise pollution assessment are *measurement microphones* which shall not be confused with high quality studio or music microphones. Both types are committed to low distortion levels since the incoming sound wave should be correctly registered by both of them. No one is expecting that the singer's voice will be distorted by the microphone, nor shall the existing sound field under measurement be distorted.

Nevertheless the greatest difference between these microphones is related to the, long term, stability regarding level perception or sensitivity. If one measurement is made showing a level of say 80dB(A) and it is repeated some weeks or months later, and the same level exists, one shall expect that the result will be the same 80dB(A). This is of fundamental importance for the comparison of different situations and for a correct enforcement of laws and regulations, for they may impose some penalties for those causing high noise levels, thus high pollution impact. The studio microphone although being of good quality can be compensated for difference in its sensitivity by simply adjusting the volume control of the amplifiers or recording equipment. The levels being, in these applications, a simple matter of subjectivity. Therefore they are not suitable for actual sound pressure level measurements.

Two major types of measurement microphone are in use: condenser microphones and electrect microphones (Fig. 3). Condenser microphones are more accurate, sensible, stable and exhibit higher sensitivity and shall therefore be preferred. They are of course more expensive than the electrect ones, which in turn are more reliable and still possess good measurement qualities (Webster, 1999).

Condenser microphones, besides the own housing are made of a diaphragm, a backplate, and an insulator. The diaphragm and the backplate create a capacitor with air as dielectric element, often polarised from an external voltage supply provided by the meter or analyser. The sensitivity of the microphone is usually linearly related to the voltage of this external polarisation. Some models of condenser microphones have pre-polarised backplates and shall not be used with the external polarisation. The diaphragm, which is quite delicate, is protected by a removable grid. This grid is intended to be part of the sensor and is only removed for metrological calibration purposes. With the pressure fluctuations the diaphragm is excited by a fluctuating force thus varying the capacitance between diaphragm and backplate, generating the output voltage of the microphone.

Figure 4 presents a typical example of a condenser microphone.

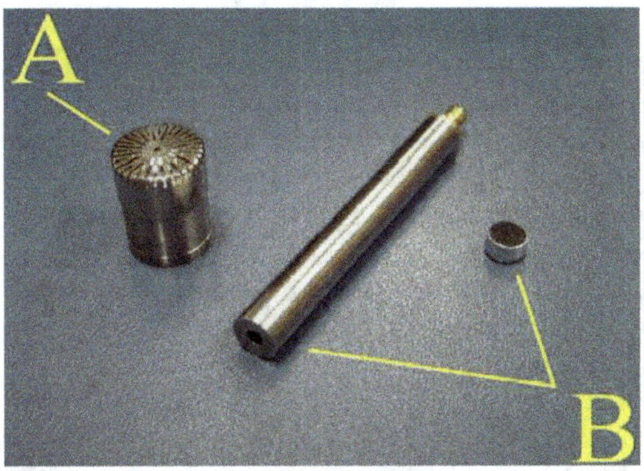

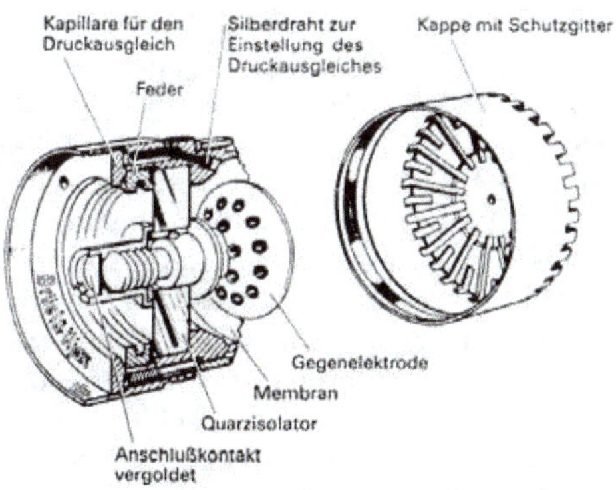

Fig. 3 and 4. Condenser (A) and electrect (B) types of microphone, and condenser microphone according to Klingenberg (1991)

The microphone characteristics, including its sensitivity, are determined by its mechanical parameters such as mass, diaphragm stiffness, tension and area, internal damping, air volume of the housing, etc. Typical sensitivities of condenser microphones range from

20mV/Pa up to 100mV/Pa. A commonly used 1/2″ condenser microphone will exhibit a sensitivity of about 50mV/Pa. This means that, when exposed to pressure fluctuations of 1Pa i.e. 94dB, the output of the microphone will be 50mV. Accordingly, since we are using a logarithmic dB scale for the SPL, with 114dB the output will be of 500mV and with 74dB of only 5mV. It is clear that the loudest and quietest levels capable of being measured will depend on the electrical measurement range of the meter but also on the microphone sensitivity. The quietest level is influenced by the electrical background noise on the meter, including cabling. For the commonly used microphones this level is about 15dB to 20dB.

The frequency range also affected by the above mentioned parameters will usually starts about 20Hz and extend up to 20kHz, since it should follow the human ear characteristics.

Since, when dealing with noise pollution in urban areas, the specified microphone shall be of *free-field* type. This kind of microphone is designed to have its frequency response compensating the presence of the microphone itself on the sound field, distorting the waves being measured. The free-field is an idealisation of a sound-field free of reflections, where the acoustic energy being measured is coming only directly from the sound source. The free-field microphones shall be pointed directly towards the sound source under investigation for they are designed to operate this way.

Condenser microphones are very sensible sensor, therefore great care shall be taken when manipulating and storing them. The protective grid shall remain in place except when it is being metrologically calibrated. This procedure consists of the verification of the sensitivity and frequency response of the microphone, done in a for this purpose accredited laboratory, and shall be repeated every two years or according to the requirements stated in local regulations, standards or applicable laws.

Electrect microphones also exhibits a diaphragm and a backplate working quite with the same varying capacitance principle. However either the diaphragm or the backplate are made of a metal plated polymer which is electrically charged to create polarisation voltage. The diaphragm built in this way usually is not suited to be tensioned in order to exhibit a similar frequency response as in the case of the condenser microphone. A better construction will use the backplate as the pre-polarised surface of the capacitor, thus allowing for a better diaphragm. The sensitivities usually found in electrect microphones range from 5mV/Pa to 15mV/Pa, thus being fair less sensitive than the condenser type. It means that the quietest levels that can be measured with electrect microphones are higher than those from the condenser type, being of about 30dB. Two great advantages of the electrect type are the reliability, the microphone itself being very robust, and the possibility of mass production, decreasing the final price of the sensor.

In both cases, with electrect as well as with condenser microphones, the pressure fluctuation causes a capacitance variation. Nevertheless the electrical circuit of a meter or analyser is usually designed to measure voltage, not capacitance. In order to transform a capacitance variation in varying voltage and also to adjust the electrical impedance of the sensor to that from the meter a preamplifier is needed. Usually, the preamplifier does not affect the sensitivity of the microphone and is designed to have a flat frequency response extending beyond that of the sensor. It shall not, therefore, significantly alter the measurement characteristics of the sensor but shall allow for a better transmission of the electric signals to the sound level meter, including eventually the cabling.

For the placement of the preamplifier should be as close as possible to the sensor they are usually the mounting point of the microphone and therefore have matching dimensions

(Fig. 5). Microphones of different diameters must use different preamplifiers or specific adapters.

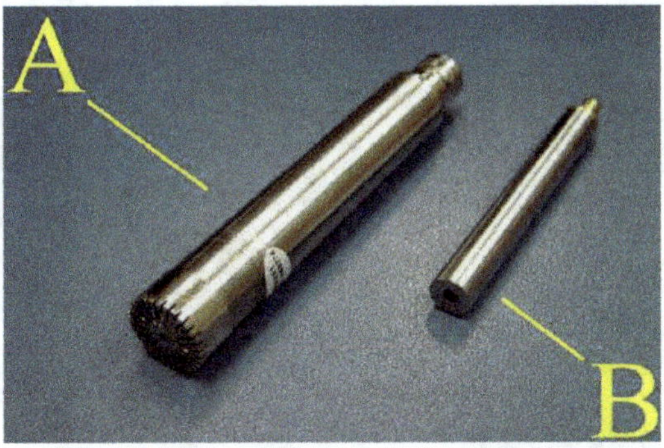

Fig. 5. Preamplifier for condenser (A) and electrect (B) microphones (microphones attached)

The signal conditioning for the electrect microphones is usually incorporated in the sound level meter and no preamplifier is usually seen, although it exists inside the meter. In some cases, where constant voltage and varying current measurement circuits are used, as in the case of IEPE interfaces, even the electrect microphones are fitted with specific preamplifiers. Due to the nature of the circuitry involved in these cases only pre-polarised microphones can be used.

The complete measurement chain consists of microphone, preamplifier, cables and sound level meter. Although being part of the chain the preamplifiers are not normally required to be metrologically calibrated, only the microphones and SLM. However, during the normal calibration or verification prior to the measurements the whole chain, including the preamplifier, must be verified with an acoustic calibrator.

3.2 Meters and Analysers

Once the pressure fluctuations were transformed into electrical signals, through the microphone and preamplifier, a device is needed to properly execute the frequency weighting, the RMS and level calculations, displaying at the end of the process the obtained values. Since the sound level is the quantity being measured one speaks of a *Sound Level Meter (SLM)*. If the device is capable of doing some kind of frequency analysis it is in turn called *analyser*, or *spectral analyser*.

If only the instantaneous values of the detectors, slow, fast and/or impulsive are available one may speak of a simple SLM. If it is possible to configure the equipment to measure an equivalent level during some specified time interval, it is then called *Integrating Sound Level Meter* (Fig. 6).

Besides the possibility of displaying the already mentioned levels from each detector the SLM shall allow the calibration of the measurement chain, with the help of an acoustic calibrator.

An international standard, IEC61672 (IEC, 2002), establishes tolerance classes for this kind of equipment, namely types 0, 1, 2 and 3. A device of type 0 shall meet the most stringent tolerances, being more accurate than the other classes. It is mainly intended for laboratory devices operating under stable controlled conditions. Type 3 devices are seldom used for noise pollution assessment since they are not accurate enough, in general use of a type 1 or type 2 device is enforced by law, regulations or standards. Equipment of type 1 are quite accurate and fitted with high quality condenser microphones. Type 2 in general use electrect microphones and have lower cost.

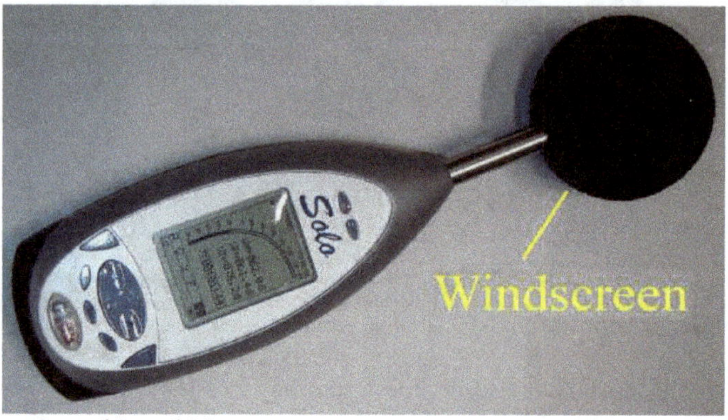

Fig. 6. Integrating Sound Level Meter

Among other characteristics it is advisable to specify and use SLM with the capability of performing statistical analysis and great storage memory, allowing a deeper insight on the time variations of the sound level during a longer period of study. Integrating SLM are mandatory for the most regulations demand the measurement of an equivalent level, in dB(A), *LAeq*.

Very useful, when dealing with noise pollution from different kinds of sound sources in the urban environment, is the assessment of the statistical percentile levels. These values, represent the acoustic level found to be surpassed during a specified percentage of the time, from an acoustic long time assessment. For instance, if the varying levels during a measurement are above a certain value, in decibels, during ten percent of the time, this value is called the Ln10, accordingly the value for which the levels are over for half of the time being considered (50%) is called Ln50 and so on.

If its difficult to measure the background noise for the sources cannot be turned off, the Ln90 can be used as an estimate. During noise pollution surveys it is usual to establish at least the percentile levels Ln10, Ln50, Ln90. The difference among them is related to the variations found in the noise impact. If one is dealing with a source operating continuously and in steady state, like an industrial equipment for instance, the percentile levels would be close to each other, a few dB, for the level itself will not vary along the measurement. With traffic noise or noise from entertainment activities for instance one will face spread percentile levels for the sound source varies with time. The long term time structure of the noise can be assessed by the percentile levels. They need to be determined from a continuous statistical assessment during the measurement and is automatically done by the sound level meters or analysers capable of doing this kind of analysis. Figures 7 and 8 shows two different situations where the Fast level is shown together with percentiles Ln10, Ln50 and Ln90, please note the different time scales in the figures. One can easily see that the difference between the percentile levels is quite smaller for the steady state noise from a turbine than in a sport stadium with people cheering.

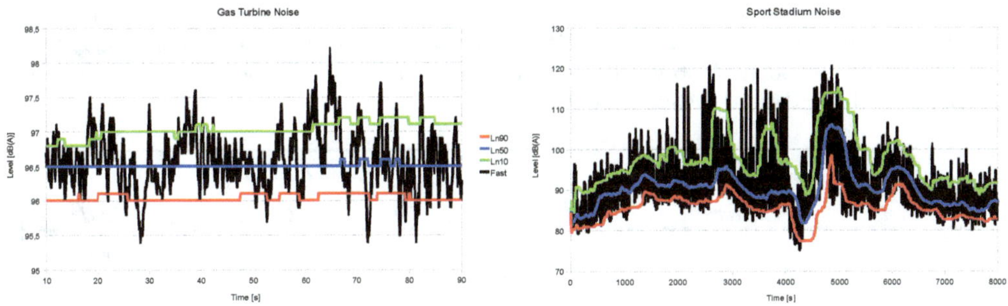

Fig. 7 and 8. Time evolution of percentile Ln levels from a running gas turbine, and time evolution of percentile Ln levels from noise in a sports arena

When complex situations are faced an efficient design of mitigation measures must take into account the nature of the sound field involved. This may lead to the necessity of determining the distribution of the sound energy among the different frequencies of the audible spectrum, in order to specify the correct absorptive materials or isolation. This is accomplished by a spectral analyser using variable width, broad band, frequency filters like the octave filters or third octave filters.

Octave filters were developed based on the human ear characteristics which tends to render similar sounds with doubling frequency ratios. A sound with 440Hz or with 880Hz are quite different from each other in the sense of frequency of the pressure fluctuations, nevertheless, both are recognised by the listener as being the same musical note. An octave is thus a frequency interval corresponding to a doubling, or halving, of some basis frequency. For the octave filters, which are standardised, the basis frequency is 1000Hz, used as the centre frequency of one of the filters. As one can easily see, the bandwidth of the filters must grow with growing frequency, obeying a power law, as in equation (3).

$$f_{i+n} = f_i \times 2^n \,. \tag{3}$$

where f_7 is the basis frequency of 1000Hz and i varies from 1 to 11, f_i varying from approximately 16Hz to 16kHz.

The initial, center, and end frequencies of the octave filters are shown on Table 1. The power law is however slightly modified to round up some frequencies.

The human ear performs a frequency analysis in the inner ear. Due to the spatial characteristics of the cochlea the frequency evaluation takes place in a way resembling filters obeying a power law similar to equation (3), as in equation (4).

$$f_{i+n} = f_i \times 2^{\frac{n}{3}} \,. \tag{4}$$

where f_{20} is the basis frequency of 1000Hz and i varies from 1 to 33, f_i varying from approximately 12.5Hz to 20kHz.

Since the exponent is now 1/3 of the exponent from the octaves these filters are known as third octave filters. One octave corresponds to three, sequential, third octave filters. Their initial, center, and end frequencies are also shown on Table 1, and are in the same way rounded. It is a rather good approximation of the behaviour of the inner ear, specially for frequencies above 150Hz. Below this frequency the bandwidth of the inner ear filtering is broader than the first third octave filters.

Octave			Third Octave		
Initial	*Center*	*End*	*Initial*	*Center*	*End*
			11,2	12,5	14,1
11	16	22	14,1	16	17,8
			17,8	20	22,4
			22,4	25	28,2
22	31,5	44	28,2	31,5	35,5
			35,5	40	44,7
			44,7	50	56,2
44	63	88	56,2	63	70,8
			70,8	80	89,1
			89,1	100	112
88	125	177	112	125	141
			141	160	178
			178	200	224
177	250	355	224	250	282
			282	315	355
			355	400	447
355	500	710	447	500	562
			562	630	708
			708	800	891
710	1000	1420	891	1000	1122
			1122	1250	1413
			1413	1600	1778
1420	2000	2840	1778	2000	2239
			2239	2500	2818
			2818	3150	3548
2840	4000	5680	3548	4000	4467
			4467	5000	5623
			5623	6300	7079
5680	8000	11360	7079	8000	8913
			8913	1000	11220
			11220	12220	14130
11360	16000	22720	14130	16000	17780
			17780	20000	22390

Table 1. Initial, Center and End frequencies from octave and third octave filters in Hz

The frequency analysis consists in the separation of the energy contents of the sound signal into distinct parts, between the initial and end frequencies of each filter. The sum of the energies of all filters will correspond to the energy of the whole signal. The frequency characteristics of the octave and third octave filters are standardised by IEC 1260 (IEC, 1995).

Once the energy is divided among the filters the same procedure is done to obtain the corresponding sound pressure levels in each frequency, including eventually the frequency weighting. From the obtained spectra, we can approximately convert from dB to dB(A) or any other weighting by just adding the values in dB shown in Fig. 2 for each weighting found at the corresponding center frequency of each filter. Please note that the majority of the values are negative, thus reducing the corresponding levels. From weighted spectra, we can find the approximate flat or linear spectra by just doing the subtraction.

As noted before the bandwidth of the third octave filters is quite narrow in the low frequency range but rather broad in higher frequencies. If there is a need to determine precisely the frequency contents of the sound, for instance to correctly identify a source or to optimise the acoustic treatment another possibility is the use of constant bandwidth filters implemented through Fast Fourier Transform (FFT) algorithms. The mathematics of the FFT is beyond the scope of this chapter but consists on the use of the properties of the Fourier transforms, actually of the Fourier series, to split the energy of the signal among several bins each with the same bandwidth, over the whole frequency range.

Since it is accomplished by a mathematical algorithm, the FFT analyser requires the digitalisation of the signal and a computer to perform the calculations. Of course the frequency analyser itself is a dedicated computer. Modern systems tends to split the functions requiring a device for the data acquisition and digitalisation and a computer, normally a laptop, with software to perform the calculations and display. The same procedure for weighted spectra conversion can be used here.
In the case of noise pollution assessment, most regulations and standards foresee a kind of penalty for the case of the presence of pure tones in the sound. This is due to the quite annoying characteristics f this kind of noise. For two distinct situations with the same overall sound pressure level, one with broad band noise, the other with the presence of a prominent pure tone, even in the context of a broad band noise, the human ear tends to regard the former as being more annoying. In order to detect and document the tonality of the noise its spectral characteristics is taken into account. Normally if any of the filters, in a third octave spectrum, or any of the bins in a FFT-analysis exhibits a level which is 5dB higher than those of its neighbours, filters or bins, the noise may be regarded as being tonal.

For more complete analysis, specially for long term monitoring, these functionalities can be combined with data transmission on a monitoring station. They may be installed on a fixed position, measuring and storing data which are transmitted to a central through, telephone lines, cell phone technology or even wireless internet connections. Different protocols are used for this data transmission, mostly proprietary protocols from the equipment producer. Monitoring stations are normally offered with a complete package including software for the data transmission, and a database to store and analyse the obtained levels.
Figure 9 presents the results from different analysis of the same signal.

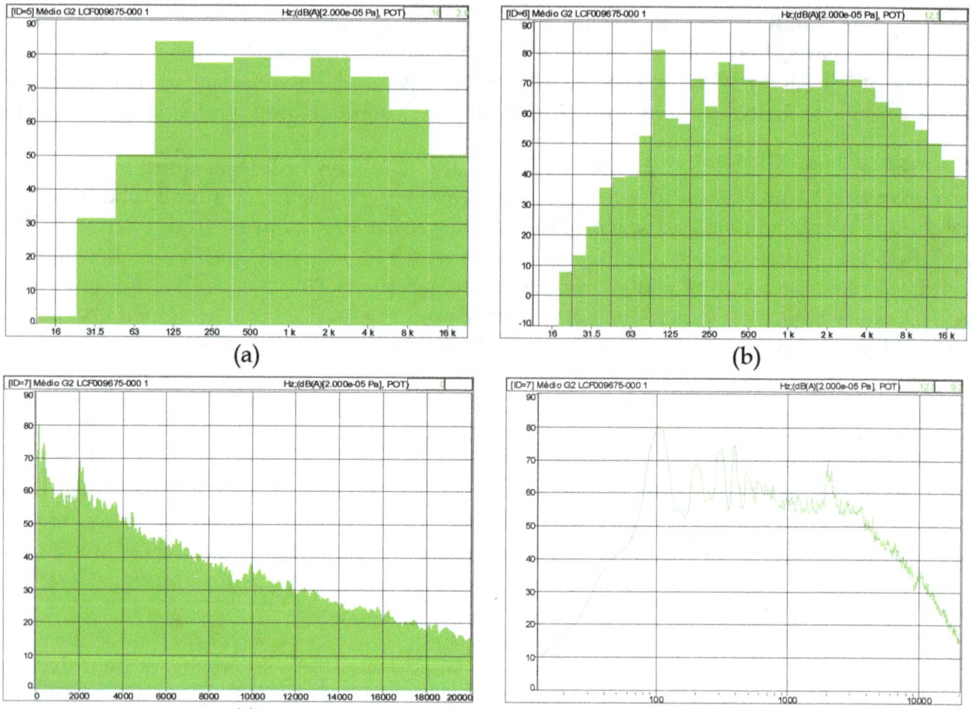

(a) (b) (c) (d)

Fig. 9. Different analysis of the same signal (Level [dB] x Frequency [Hz]): (a) Octave, (b) Third Octave, (c) FFT Linear freq. Axis, and (d) FFT Logarithmic freq. Axis

3.3 Choice of Detector and Time Interval

Usually, during a noise pollution assessment, one will deal with time varying sound pressure levels. In order to evaluate the measurements it must be distinguished among equivalent levels over a time period and the instantaneous levels. For the sake of allowing a comparison between different measurements, done with different equipment a set of detectors were defined in the standard IEC 61672 (IEC, 2002) which will provide the instantaneous level of the measurement. They reflect the behaviour of the acoustic signal in different time scales, from 1 second to a few milliseconds. Although they may be actually regarded as equivalent levels for this time interval, calculated according to equation (1) with the corresponding value for T from the Table 2, the duration is quite small when compared with the time scale of a general assessment of noise that they may be considered as instantaneous. Table 2 shows the different standardised detectors and their time scales.

Detector	*Averaging time*	*Use*
Slow	1 s	Outdoor noise
Fast	0.125 s (1/8 s)	Outdoor and indoor noise
Impulse	0.032 s with longer decay time	Short duration impulsive events
Peak	Instantaneous	Impulsive noise

Table 2. Sound level meter detectors

If a graphic is plotted with these values, we will have the time history of the noise being measured. Of course the Fast detector will better follow the fast variations of the signal. Actually the regulation or standards for noise assessment being followed will determine the kind of detector to be used. The Impulsive detector is more appropriate for the evaluation of impact noise, as well as the Peak detector. The Slow and Fast are more used in urban noise assessment. Figure 10 shows the time history of the same acoustic event measured with different detectors.

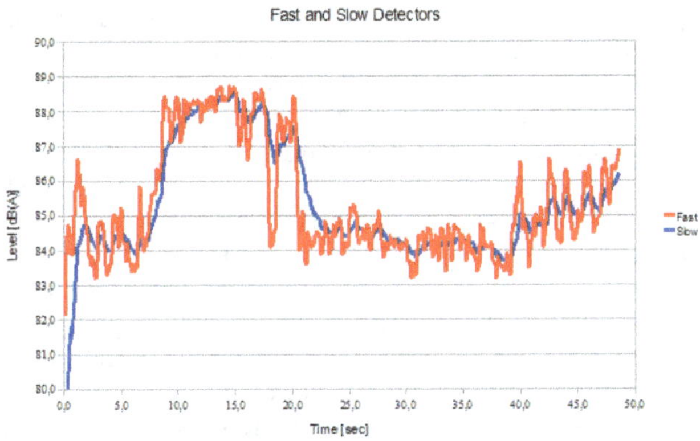

Fig. 10. Time history of a signal with different detectors

With the advent of the integrating sound level meters the parameter for a long term evaluation of urban noise changed for the equivalent level *Leq*, in most cases with A-weighting the *LAeq*. The evaluation is done again according to equation (2) but the choice of the averaging time *T* shall reflect the behaviour of the noise being assessed. The averaging time must thus be chosen in order to reflect the main variations of the sound, reflecting a representative portion of its overall behaviour. In the case of a traffic noise for instance, it must cover periods of red lights as well as periods with normal traffic. A time span of 5 minutes is normally a good choice. For industrial noise, with equipment running continuously and in steady-state about 1 minute may be a good choice. In the same way one can build a time history of the sound with *Leq* levels or also build a time evolution of the *Leq* parameter allowing the time *T* to continuously increase until it eventually reach a constant level. This will correspond to a general average of the acoustic energy over this period. Figure 11 shows the time history of 5 minutes *LAeq* from traffic noise over a whole day in two different locations. For comparison, the instantaneous Fast levels are also shown.

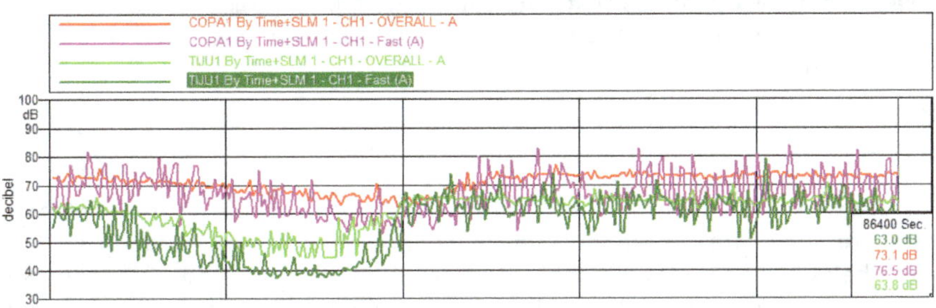

Fig. 11. Time history from Traffic noise in a 24h period - *LAeq* 5 min. and Fast

Of course when stating a *Leq* level, we must also mentioned the time period used for its determination.

3.4 Acoustic Calibrators

The primary objective of the acoustic calibrator is to provide a reliable reference of frequency and amplitude for the pressure fluctuation. Thus it allows the quick verification of the quality and accuracy of the measurement chain, from the microphone to the meter or analyser (IEC, 2003). It consists of an equipment capable of generating a simple harmonic acoustic noise signal , with known amplitude and frequency.
Usually the amplitude values correspond to either 94dB or 114dB, 1Pa or 10Pa respectively. The use of higher values is related to a better stability of the electro-mechanical mechanism when compared to the generation of a relatively small pressure of 1 Pa.
The frequencies most commonly found in calibrators are 250Hz and 1kHz. Whenever possible use shall be made of a frequency of 1kHz. Since every standard weighting filter, from A to D, do not impose a correction factor at 1kHz, the imposed amplitude of the harmonic sound wave value is expressed as the same SPL, either 94dB or 114dB depending on the choice, in whatever filter chosen, and even without weighting (dB(F) or (L)). This facilitates the setup of the measuring equipment, reducing the chance of an operator's error.
Since the conversion of pressure fluctuation into electric signals is mainly determined by the microphone sensitivity, the measurement equipment must be properly set to correctly convert the voltage received to pressure. This can be achieved by exposing the microphone to the known pressure generated at the calibrator output, setting up the measurement either to the global sound pressure value or to the specific frequency of the calibrator, and observing the instruments reading. Using its setup options the obtained value shall be corrected to correspond to the calibrator's level, thus assuring the accuracy of the measurement chain. This procedure shall be repeated before and after each measurement set and is of course mandatory after a microphone replacement, for its sensitivity will likely be different from the previous one.
After the measurement set is finished, the calibrator shall be used again to verify if any changes have occurred in this period. The difference obtained between the calibrator level and the reading at the end of the measurement set is an indicator of the smaller bias error of the values obtained. Of course other factors affecting the measurement will still need to be added upon this bias.
Care should be taken when positioning the microphone in the calibrator's opening for it must fit properly in place. Otherwise the imposed pressure on the microphone will not correspond to the expected level. The volume of cavity formed at the calibrator, with the microphone closing the opening, is extremely important, therefore.
For different standardised microphone diameters one may find suitable adapters for the calibrator's opening.

4. Positioning

A very important issue, of course, is related to the place where the measurement takes place. Sound pressure levels are consequence of an existing sound field build up by the sound waves and their interactions with the environment. Reflection and diffraction are very important factors to be considered as well as the eventual interference between sound waves coming from different sources or reflections. Much like a temperature field, it is hotter near a heat source such as an oven than far away from it, the sound pressure level may vary from place to place. Therefore the actual position where the measurement is being made will influence the result.

In the urban noise assessment of primary concern is the position where the population is exposed to the pollution. Although a measurement near the sound source may be very useful to characterise this source the population may be exposed to different levels in different places. For instance, for a source like traffic noise, there may be remarkable differences between levels measured at the side walk, at the first floor in a building, at the highest floor in the building and even in windows facing directly to the street or facing the back of the building. Thus together with the sound pressure level one must also report the exact location where it was obtained. Commonly used measurement heights are 1.2m to 1.5m, corresponding more or less to the ear level and 4.0m in urban surveys, avoiding being too close to the traffic noise source. The regulations or standards followed shall indicate the measurement height when applicable.

A different situation arises when dealing with specific complaints. In these cases preference should be given to measurements in the actual places where the complaints are made. Some regulations however (ABNT, 2000, and City of Rio de Janeiro, 1978, 1985, 2002) allows the measurements to be done from a specified distance from the sound source. Sound source in this case, from the point of view of urban legislation, may be regarded as being the terrain or property where the sound is being generated, where the emission comes from. It is not the equipment, machine, sound system or whatsoever which is regarded as source since one will not have, in general , permission to enter the property to evaluate the noise. Therefore the property itself is considered as source.

When measuring inside of rooms care should be taken with the eventual presence of standing waves due to resonance. Usually at least three measurements are done and their average value is taken as the result. Positions close to walls or furniture must be avoided, with a distance of at least 1m. The values measured will also differ if measured directly on the facade, or with open or closed windows. The actual position must be correctly reported to allow a further verification or comparison to other evaluated levels.

Avoided in any case must be measurements too close, less than 1m, to reflective surfaces of any kind such as walls, vehicles, trees, etc... Barriers shall also be avoided, including here even the very presence of the operator of the sound level meter in the sound field.

No measurement shall be done under rain conditions, for the rain noise in general is not considered in the allowable limits and the high humidity is dangerous to the measurement equipment. If the wind speed is above 5m/s the measurements may be affected by the wind noise itself, generated by the presence of the microphone in the air flow. Without measurement there will be no floe disturbance and the wind noise measured will not exist. Therefore measurements under these conditions are not reliable. In the presence of wind, with lower speeds, a suitable wind protection, windscreen, must be used to reduce the flow noise induced at the microphone diaphragm.

5. Conclusion

The correct assessment of noise levels starts with the measurement of the physical stimulus pressure fluctuations in order to quantify the sensation caused, the sound. In order to perform actual measurements which may be useful to enforce the application of laws and regulations a thorough knowledge of the characteristics of the involved sensors and equipment as well as the correct choice of configuration parameters. Among others a suitable time detector, frequency weighting filter, correct calibration and bandwidth of the spectral analysis are of primary importance. These aspects, together with the positioning of the microphones and the evaluation of the commonly used metrics were discussed to help the choice of measurement configuration in order to understand the environmental impact of noise pollution in urban centres.

6. References

ABNT - Assosiação Brasileira de Normas Técnicas (2000). *NBR 10151/2000 Acústica - Avaliação do ruído em áreas habitadas, visando o conforto da comunidade - Procedimento,* Rio de Janeiro, Brazil

City of Rio de Janeiro (1978). *Decree #1,601 from June 21st (1978),* Diário Oficial do Município do Rio de Janeiro, Brazil

City of Rio de Janeiro (1985). *Decree #5,412 from October 24th (1985),* Diário Oficial do Município do Rio de Janeiro, Brazil

City of Rio de Janeiro (2002). R*esolution #198 from February 22nd 2002 of the environmental board of the city,* Diário Oficial do Município do Rio de Janeiro, Brazil

IEC (1995). *IEC 1260 Electroacoustics - Octave-band and fractional-octave-band filters,* International Eletrotechnical Comission, Genève, Switzerland

IEC (2002). *IEC 61672-1 Electroacoustics - Sound Level Meters Part 1 Specifications,* International Eletrotechnical Comission, Genève, Switzerland

IEC (2003). *IEC 60942 Electroacoustics - Sound Calibrators,* International Eletrotechnical Comission, Genève, Switzerland

ISO (1975). *ISO 532 Method for Calculating Loudness Level,* International Organization for Standardization

Klingenberg, H. (1991). *Automobilmeßtechnik Band A: Akustik,* Springer Verlag, Berlin

Schick, A. (2004). History of psychoacoustics, In: *Proceedings of 18th International Congress on Acoustics (ICA),* Kyoto Acoustical Science and Technology for Quality of Life, Vol. V, pp. 3759-3762

Webster, J.G. (1999). *Measurement, Instrumentation, and Sensors Handbook CRCnetBASE,* CD-ROM, CRC-Press, Boca Raton, USA

11

Experiences with the Urbanisation of Slums: Management and Intervention Models

Adauto Lucio Cardoso, Angela Maria Gabriella Rossi
Federal University of Rio de Janeiro (UFRJ)
cardoso@ippur.ufrj.br, gabriella.rossi@poli.ufrj.br
Brazil

1. Introduction

Brazilian municipalities have accumulated a substantial amount of administrative and technical knowledge about the methods and alternatives for offering technical assistance in precarious settlements. So far, however, that experience is still fragmented. It is thus necessary to identify and consider both the successes and the failures that have been made.

A main controversy on the topic involves the question of whether to establish a national policy, managed centrally by the Ministry of Cities (*Ministério das Cidades*), to establish technical and environmental quality criteria in the activities to be carried out under an Urban Plan, or to not have such criteria and instead determine the criteria for each individual case at the local level. Considering that the resources used for such projects are public and federal, and that the responsibility for implementing these plans lies with the responsible body, there is a strong argument that some kind of control should be established so that the managing party can be made accountable for the results. This would require comparing the results with some established criteria.

Brazilian historical experience does not recommend the adoption of general models or patterns for the interventions, due to the risk of creating insurmountable regulatory barriers and recognising that the residents of the precarious settlements have a right to housing. The regional inequalities and the different situations that characterise the various existing settlements reinforce the difficulty of establishing universal standards, which may result in criteria that are not adequately suited to local conditions and needs.

Another argument used to promote the need for standards is related to the planners' aesthetic and ideological tendencies, which are marked by rationalism and the modernist functionalism that do not recognise the environmental and aesthetic virtues of these settlements.

A radical opinion is that interventions in the slums and other precarious settlements should interfere as little as possible with their urban structure, respecting the physical characteristics of the settlement and the right of people to remain in it, and limiting radical changes to those settlements such that the only changes are necessary to solve serious problems or to provide growth or extensions of roads to ensure accessibility. This has been the decision of many municipal programmes that have dealt with the urbanisation of slums.

Other arguments can be considered as well. One of these opinions considers the problem of violence in the slums and the importance of restructuring the spaces as a way to ensure democracy and public use.

The ways of structuring the precarious settlements do not necessarily constitute a choice or "a construction project" but, rather, an extremely desperate need for which there are no other options at hand.

Another important point concerns the establishment of "minimum standards". The debate initiated by the rationalism in architecture and engineering at the beginning of the twentieth century can be seen from two viewpoints. On one side, by proposing minimum standards, regulatory controls on the market can be established to avoid the diseases and epidemics often present due to "non-hygienic" houses. This was meant to establish the principles of a new type of citizenship, guided not only by personal and political rights but also by social rights. On the other side, it was considered by some to be a mechanism of control and domestication of poor people in order to guarantee the dominion of capital.

The principles of rationalism made *tabula rasa* of cultural realities and needs, creating housing designs that followed the mandates of rationalisation of space and buildings, on one side, and the cultural options of the middle classes, seen as universal needs, on the other. The inadequacy of that model for the care of the real needs of the grassroots was emphasised by the criticism of John Turner (Turner, 1972) and, in the case of Brazil, Carlos Nelson Ferreira dos Santos (Santos, 1977). An extreme example was the removal of slums that occurred during the military dictatorship.

The problem is that the criticism - necessary and appropriate - of modern rationalism obscures its positive aspects, particularly the idea that each citizen has the right to decent housing: healthy, safe, functional, and in a sustainable environment, with adequate access to public spaces, infrastructure, and collective equipment. Certainly, nowadays everyone agrees with this. The question is how to make the concept of decent housing operational in programs of urbanisation of precarious settlements.

Recent Brazilian experiences have shown great progress in several areas, including the development of methods of intervention, experienced technical teams, diagnosis techniques, and improved designs. A disparity concerns the degree of intervention to reduce population density, either with structural improvements in the settlements or by relocating homes in the situation of extreme hazards.

Some projects have set certain standards, such as for the width of streets, minimum lot sizes, the maximum length of a street, the distance between steps in staircases, and so on. Other interventions have been limited to solve more basic issues: areas of risk, minimum accessibility, common areas, infrastructure, and some collective equipment. There are, in these cases, situations of the consolidation of homes in extremely precarious situations: houses made with improvised materials, buildings at risk of collapse, narrow alleys between tall buildings where the sun never penetrates, flooded land, or blocks with internal houses where it is not possible to open windows without being completely surrounded by other buildings.

Based on the above arguments, it is recommended to work towards the establishment of flexible standards, adaptable to regional conditions, that are gradual and generic enough to avoid impediments to the process of regularisation and that are able to respond at the same time to the legitimate need of respecting differences and ensuring decent housing.

This chapter seeks to identify the basic points and some very general criteria that could be used as parameters for intervention and design.

The first consideration relates to the type of intervention. The accumulated experience allows us to identify 3 basic types: urbanisation, re-urbanisation, and removal. Urbanisation means that the intervention does not modify the structure of the settlement, but rather only the building infrastructure and paving. It is the appropriate intervention for cases where the densities are lower and the design of settlements is compatible with existing standards . Re-urbanisation includes interventions that, because of the inadequacy of the site, completely replace the structure of the settlement by dividing and relocating families in the same area. Removal means the removal of the population and their resettlement in another area. This may be the case when the settlement occupies an at-risk area or an area of environmental preservation of "inestimable ecological value." In practice, several interventions make use of all 3 alternatives, relocating a part of the settlement and removing a portion of the population. The point here is what the main characteristic of the intervention is.

The choice among the 3 alternatives depends on the conditions of the site and the minimum standards that are or are not used. The adoption of standards often involves the need to lower the population density, which involves the division of lots and, in some cases, the removal process.

Another way to classify the interventions relates to their scope : they can be complete or partial. That is, they can be related either to a part of the settlement (partial) or to the whole settlement (complete). Partial interventions have been historically done in slums and in lots, possibly because of clientelistic political practices or in emergency situations. The problem with complete interventions is the need for a significant amount of resources, covering the settlement as a whole, which limits action in only a few settlements. An alternative to this problem has been the adoption of gradual planned interventions, in which partial interventions are linked to a global plan of action and a technically consistent sequence of actions. Gradual planned interventions may be adopted in cases where the local government provides assistance to several settlements at the same time, which is only financially feasible by extending the time of the finalisation of services.

Another important distinction concerns the components of the intervention, allowing its characterisation as sectoral or integrated. In general, integrated projects are those that include the social dimension in addition to the physical-urban and environmental dimensions. The components of an integrated project vary according to the specific needs of the settlement and also with the available budgetary possibilities.

In general, there is a consensus between technicians and managers that complete (or gradual planned) interventions and integrated projects have greater effectiveness than partial or sectoral interventions. It is necessary, however, to remember that the adoption of one kind or another depends on the needs of each settlement and also on the available resources.

In this sense, it is recommended that the interventions should be mainly complete or gradual planned and integrated, including popular participation, land regularisation, physical-urban, social and environmental intervention, and post-occupation measures, highlighting urban regularisation. It is important to note, however, that this general recommendation should be adapted to the conditions of financing and to local needs.

For each settlement, a plan of intervention is developed based on the physical-urban, environmental, and social diagnosis. The plan must include priorities and stages of intervention, considering the technical and social needs and the available financial resources. It is essential that the plan of intervention must also consider the environmental conditions, taking as reference, for example, the elements of the urban structure. Considering the difficulties of understanding and adequately controlling all the elements necessary for the implementation of a feasible executive project, the intervention plan must include a basic project that provides the detail and the generic elements of the construction but ensuring a degree of flexibility.

Regarding quality standards, the elements that will be included as components of the integrated project should be considered. A study done by the *Instituto de Pesquisas Tecnológicas* - IPT, Denaldi (2003) defined 3 levels to describe the types of intervention according to physical-urban quality standards:

- Minimum Standard: includes networks of water, sewage, energy, drainage, solutions for the collection of waste (direct or indirect), accessibility, treatment of areas of risk (which may involve relocation or division of lots), and land regularisation;
- Intermediate Standard: in addition to the above, this also includes urban equipment, road systems, and environmental interventions in order to address environmental problems and urban structure;
- High Standard: in addition to the above, this includes adequacy of population density, division of lots, rehabilitation or reconstruction of housing units in insecure and unsanitary situations, and an adequate road system with minimum parameters of width of streets, ensuring accessibility to all homes with pedestrian ways.

The choice between these standards depends, first, on the conditions found in the settlements and local needs. In some cases, it is not required to update the road system or to lower the population density. In general, in the context of slums, the choice depends on how to solve the equation limited resources vs. decent housing. Simulations developed by IPT for the intervention of urbanisation in a slum situated on the banks of the *Represa Guarapiranga* (São Paulo) showed costs ranging between R$ 2,300.00 and R$ 26,500.00, depending on the alternative to be used. Evaluating several intervention projects in the area of the *Guarapiranga*, the IPT found costs ranged from R$ 4,000.00 to R$ 11,000.00 per family. One of the conclusions of this study was that the adoption of different standards (in terms of average density, lots, and road system) generated cost differences averaging around 30%, considering that these are slums in the same context and under the same program. Thus, considering the scarcity of resources, the adoption of higher standards implies a higher cost, which results in being able to deal with a smaller number of settlements.

Another issue to be considered for the choice of standards is the local culture and the historical ownership of land. In this sense, some cities have far fewer problems with population density than cities like Rio de Janeiro and São Paulo, which are characterised by the scarcity of land. This element points toward a regional differentiation of patterns.

A central element to be considered is the popular participation in all spheres of deliberation. It is crucial that, in addition to the forums to be established through the creation of Municipal and State City Councils, it is important to create forums for the participation and social control of operations involving the groups directly affected. In this respect, interventions must be accompanied by commissions or committees composed of representatives chosen freely by the communities, which must be involved from the stage of drafting the plan of intervention until the post-occupation.

Intervention projects are developed to enable broad understanding by the population, even considering the different alternatives with respect to the adoption of certain quality standards and their impact on cost and the final quality of the project. In addition, in terms of monitoring, it is essential that the local committee monitors the implementation schedule and oversees the development and quality of services, with the authority to stop any actions involving a waste of resources or misuse of the project. In the final evaluation of the intervention, a separate report should be prepared by the local commission.

One of the essential elements of the integrated project is the social component, which involves the support for children and youth, generation of income, job training, and so on. For a better result, it is crucial to develop a project that take into account the needs and potentials of the community. Besides the social-economic data, which allows portraying the lives of families, it is important to assess the social-political conditions of the community, identifying resources and leadership through the use of participatory methodologies.

The social project has to be developed by a qualified team, integrated with the team that focuses on the physical-urban aspects, in order to solve problems of relationships with the community that usually arise in the course of the intervention. The social project must come before the physical design and should remain after the end of construction as an important part of the post-occupation activities.

Experience has shown that the post-occupation period is essential to ensure the objective of integrating the regularised settlements into the city. The post-occupation activities comprise the permanent social project, training courses, and generating income, as well as other activities related to the physical and urban dimension.

An important element is to check the suitability of the proposed solutions to the needs of the residents in the new context, as well as the quality of the work and the proper functioning of the infrastructure and equipment installed. In that sense, the post-occupancy evaluation is an important instrument for measuring the compatibility of solutions able to correct the problems, and it produces important data to guide new operations.

Another key element is the establishment of a pattern of urban control, defined from the existing situation during the project and after its conclusion, which is widely discussed and negotiated with the population. It must be considered that there are often informal practices, and it is slow and intensive work to create awareness of the importance of urban rules that represent limitations on individual actions. It is essential that these rules are simple, setting the spaces for public use and establishing some control of population density.

With regularisation and urbanisation, the costs of housing can rise, particularly those relating to taxes and tariffs of the concessionaires of public services. Special conditions and reduced tariffs should be negotiated for the population. The Statute of the City (*Estatuto das Cidades*) provides for the use of lower tariffs, which may also be included in the Master Plan or in specific legislation. In the case of services controlled by the state or private companies, the solution often requires extensive negotiations that, if mediated by the municipal government, will have more chances to be successful (Bassul, 2005).

2. Some Experiences of Slums Upgrading

Some of the precarious settlements that proliferate in large and medium-sized Brazilian cities have been the focus of urbanisation projects, which have been based on different models according to the region and to the public administration which manage them.

This part of the chapter shows the results of a recent two-year study done by four Brazilian Universities and financed by *Financiadora de Estudos e Projetos* – FINEP, of the *Ministério da Ciência e Tecnologia*. The research described and analysed several experiences that took place in four Brazilian states: São Paulo, Rio de Janeiro, Minas Gerais, and Bahia. Table 1 shows the analysed case studies.

State	Program	Name of the Case Studies	Public Institution that promoted the Intervention
SÃO PAULO (SP)	Saneamento Ambiental da Bacia do Guarapiranga	Sete de Setembro	Prefeitura Municipal de São Paulo
		Parque Amélia Santa Margarida	
	Urbanização Integral	Parque Capuava	Prefeitura de Santo André
RIO DE JANEIRO (RJ)	Favela-Bairro (Conde & Magalhães, 2004)	Morro da Providência	Prefeitura da Cidade do Rio de Janeiro
		Vila Mangueiral	
		Vigário Geral	
MINAS GERAIS (MG)	Plano Global Específico (URBEL, 2002)	Vila Senhor dos Passos	Companhia Urbanizadora de Belo Horizonte URBEL
		Vila Nossa Senhora do Rosário	
	-	Bairro Córrego da Ilha	Prefeitura de Sabará
BAHIA (BA)	Ribeira Azul	Group I	Companhia de Desenvolvimento Urbano do Estado da Bahia CONDER
		Group II	
		Group III	

Table 1. Case Studies (FINEP, 2007)

The results of the research are divided into three groups:

- analysis of the costs of urbanisation (Lareau, 2005);
- analysis of the Index of Environmental Health of Urbanised Slums (Almeida, 1999, and Abiko & Almeida, 2000);
- analysis of the post-occupancy

The case studies from São Paulo are:

- *COMUNIDADE SETE DE SETEMBRO* (Fig. 1): situated in the São Bernardo basin, at the right margin of the *Bacia do Guarapiranga*. The community started occupying a municipal public area in 1970. The area suffered from erosion and floods. The intervention project began in October 1994 and concluded in January 1996.
- *PARQUE AMÉLIA SANTA MARGARIDA* (Fig. 2): situated in the Guavirutuba basin, at the left margin of the *Bacia do Guarapiranga*. The community occupies a municipal public area that has slopes between 20% and 30%. The intervention began in January 1998 and concluded in August 1999.
- *NÚCLEO HABITACIONAL PARQUE CAPUAVA* (Fig. 3): situated in the extreme northeast of the Municipality of Santo André, in the Metropolitan Region of São Paulo. The area has 30% slopes and the construction was not concluded as of 2005.

Fig. 1 and 2. Comunidade Sete de Setembro, and Parque SantaAméllia Santa Margarida

Fig. 3 and 4. Núcleo Habitacional Parque Capuava (SP), and Morro da Providência (RJ)

The case studies from Rio de Janeiro are:

- *MORRO DA PROVIDÊNCIA* (Fig. 4): situated on a hill and occupies the top and the north and south slopes. The area has 6,000 inhabitants and an area of 94,000 m^2. The occupation dates from the end of the nineteenth century.
- *VILA MANGUEIRAL* (Fig. 5): situated in the west, in Campo Grande, at the margin of a river. It has 5,200 inhabitants and an area of 92,018 m^2.
- *VIGÁRIO GERAL* (Fig. 6): the entire area was, until the 1930s, a large farm that was divided into small communities. It has 6,804 inhabitants and an area of 211,956 m^2.

The case studies from Minas Gerais are:

- *VILA NOSSA SENHORA DO ROSÁRIO* (Fig. 7): situated in the eastern region of the Municipality of Belo Horizonte. Occupation began in 1940.
- *VILA SENHOR DOS PASSOS* (Fig. 8): the occupation of this area began in the 1940s and 1950s, and the slopes are as steep as 40%.
- *BAIRRO CÓRREGO DA ILHA* (Fig. 9): situated on a slope in the Municipality of Sabará. The intervention is partially complete and started in the 1980s through self-help processes.

Fig. 5 and 6. Vila Mangueiral, and Vigário Geral

Fig. 7 and 8. Vila Nossa Senhora do Rosário, and Vila Senhor dos Passos

Fig. 9 and 10. Bairro Córrego da Ilha, Sabará (MG), and Group I, Salvador (BA)

The case studies from Bahia are:

- *GROUP I - Boiadeiro, Joanes Centro Oeste, Alagados I* and *Alagados II, Atlântico* (Fig. 10): situated on a slope in the Municipality of Sabará. The intervention is partially complete and began in the 1980s through self-help processes.
- *GROUP II - Alagados IV* and *V* (Fig. 11)
- *GROUP III - Nova Primavera, Joanes Azul, Araçás I* (Fig. 12)

Fig. 11 and 12. Group II, and Group III, Salvador (BA)

2.1 Analysis of the Components of Urbanisation Costs

The costs of urbanisation for the case studies were obtained from data provided mostly by the public agencies that promoted the interventions, among others. In the case study of Minas Gerais, *Bairro Córrego da Ilha*, there was no available information about the costs of the urbanisation projects.

The particularities of each settlement (in terms of differences in its social and physical characteristics) impose different priorities to be given to type of cost component, with variable costs per unit. It is also important to emphasise that cost analysis, in the case of slum upgrading experiences, is difficult, because of the quality of the available data.

This analysis was performed on a comparative way. Table 2 shows an analysis of the data, which were divided into three groups: infrastructure, superstructure, and operational activities, and Table 3 shows the total costs of each urbanisation per family, updated for December 2005.

Comparing the three case studies from São Paulo, it can be seen that the technical constraints of the different communities predominantly influenced their costs. *Parque Capuava*, from Santo André, had the lowest cost per family, probably because it was developed over a long time, trying to minimise the overall impact on families and to adapt it to the urban structure already in place and to the type of work developed by each household. Another issue that seems to have contributed to its success was the emphasis on monitoring and social development, in addition to the direct participation of many sectors of the municipal government.

The other two interventions from São Paulo, *Sete de Setembro* and *Parque Amélia Santa Margarida*, demanded more radical solutions to sanitation, involving plumbing, water courses, and, in one of these cases, a large number of removals. This emphasises that this program - *Programa de Saneamento Ambiental da Bacia do Guarapiranga* (COBRAPE,

2001) - primarily aimed to use the urbanisation of slums for a larger goal of fighting economic and environmental problems caused by degraded areas of the basin around the Guarapiranga reservoir. In general, the urbanisation achieved their goals, including obtaining satisfactory rates for the Index of Environmental Health of Urbanised Slums.

Infrastructure	Superstructure	Operational Activities
Road System	Purchase of Areas	Executive Project
Network of Supply of Water	Rental of Areas or Equipment	Construction Management
Network of Sewage	Assembly of Temporary Accommodation	Construction Supervision
Network of Drainage	Construction of Homes	Construction Maintenance
Containment	Relocating	Technical Advice
Network of Electric Energy	Reforms in Affected Houses	Social Assistance
Network of Public Lighting	Buildings	Other Items
Areas with Equipment for Collective Use	Demolition	
Collection of Garbage	Technical Services	
Urban Furniture and Signalling		
Special Unites		

Table 2. Groups of Components of Costs

State		Case Study	Infrastructure	Superstructure	Operational Activities	Total (R$/Family)*
1	SP	Sete de Setembro	12,706.00	1,320.48	5,270.25	19,296.73
2	SP	Parque Amélia Santa Margarida	9,604.67	10,992.88	6,156.53	26,754.08
3	SP	Parque Capuava	5,509.10	4,683.63	1,105.89	11,298.62
4	RJ	Morro da Providência	8,614.86	820.66	1,160.15	10,595.67
5	RJ	Vila Mangueiral	5,824.04	958.76	677.68	7,460.48
6	RJ	Vigário Geral	19,132.49	985.92	2,380.16	22,498.57
7	MG	Vila Senhor dos Passos	5,079.77	2,700.06	1,453.47	9,233.30
8	MG	Vila Nossa Senhora do Rosário	3,669.71	8,996.72	1,427.44	14,093.44
9	MG	Bairro Córrego da Ilha	-	-	-	-
10	BA	Group I	11,827.01	11,518.45	-	23,345.46
11	BA	Group II	7,044.88	2,478.08	-	9,522.96
12	BA	Group III	14,771.63	1,150.66	-	15,922.29

* US$ 1.00 approx. R$ 1.50

Table 3. Total Costs of Case Studies

In the case of Rio de Janeiro, *Vigário Geral* had the highest cost per family, which can be explained by the large size of the slum and the need for two stages of urbanisation because the value estimated for the first contract was not sufficient to finish the planned work.
In Minas Gerais, the highest costs per family were in *Vila Nossa Senhora do Rosário*, perhaps because the costs provided to the urbanisation of *Vila Senhor dos Passos* are related only to one type of program (in this case, the urbanisation was provided by two different programs: *Habitar-Brasil* and *Alvorada*, and the costs related to the latter were not available). The two communities have similar physical characteristics; however, *Vila Senhor dos Passos* has a greater aerial extent than *Vila Nossa Senhora do Rosário*.

2.2 Analysis of the Index of Environmental Health of Urbanised Slums

The Index of Environmental Health for Urbanised Slums is defined by fourteen indicators and aims to measure, in an individualised way, the elements that interfere with the functionality of existing actions in the slum. After scoring them on a scale from 0 to 100, these indicators are grouped in the calculation of an average for the Index. Urban slums are considered to have good health when they have a score above 85, moderate health when the score is between 70 and 85, and poor health when the score is less than 70. The final score may be used as a reference for the ranking of slums, which allows to the characterisation of the level of need among different slums in the same municipality.
In the case of this research, the analysis using the Index was done by comparing only those cases where the urbanisation process had already been finalised, which include the three cases from Rio de Janeiro, *Morro da Providência*, *Vila Mangueiral* and *Vigário Geral*, and the two cases from São Paulo, *Sete de Setembro* and *Parque Amélia Santa Margarida*. As the integrated data analysis depends on the integrity of the survey, the absence of some data may interfere in the conclusions, causing possible misinterpretations. Table 4 shows the scores related to the five case studies.

Case Study	INDICATORS														INDEX
	Density	Income	Education	Water	Sewage	Garbage	Drainage	Road System	Geology	Electric Energy	Sweeping	Public Lighting	Public Space	Land Regularisation	
Sete de Setembro	80	20.1	0	100	100	100	90.3	82.85	100	100	93.33	49.7	60	80	79.26
Parque Amélia / Santa Margarida	80	9.9	0	100	100	100	90.6	70.01	100	100	76.68	41.71	100	80	78.74
Morro da Providência	80	79	67	90.06	90	90	90	29.18	100	90	100	100	60	80	81.87
Vila Mangueiral	80	79	80.20	90.22	90.22	90	90	65.2	100	100	100	100	100	80	88.98
Vigário Geral	100	23	67	90	90	90	90	64.24	100	100	100	100	100	80	85.31

Table 4. Index of Environmental Health of Urbanised Slums

Only two of the cases from Rio de Janeiro, *Vila Mangueiral* and *Vigário Geral*, had positive scores. *Morro da Providência*, also in Rio de Janeiro, was classified as moderate. The two cases from Sao Paulo, *Sete de Setembro* and *Parque Amélia Santa Margarida*, have moderate scores with very similar values. The majority of the cases are in a favourable situation, but the indicators of income, education, traffic routes, public lighting, and public space are, for the most part, classified as unsatisfactory.

2.3 Post-Occupation Diagnosis

Here, the main results of the post-occupation diagnoses, made through information obtained from the respective public institutions that promoted the urbanisation processes, are presented.

If the defined objectives and results were achieved

At the time of this study, 100% of the water supply and sanitation networks and all geotechnical services and road systems had already been implemented in *Parque Capuava* (Santo André, Sao Paulo); the electric grid, street lighting, and garbage collection system had not yet been completed. According to the technical explanations, the deployment of these services is the last stage of urbanisation. In the cases of *Sete de Setembro* and *Parque Amélia Santa Margarida*, the goals and results defined by the Environmental Sanitation Program of the Guarapiranga Watershed (the program that promoted the urbanisation of these slums) were completely achieved.

In the three case studies from Rio de Janeiro, the objectives defined in the projects and results had been achieved.

In Minas Gerais, the process of urbanisation had not been fully completed at the time of this research. At *Vila Senhor dos Passos*, where the urbanisation process is more advanced, the goals and results had been achieved.

In Bahia, the works were still in progress in some cases (*Alagados IV and V, Joanes Blue*, for example), and there was no post-occupation diagnosis.

If the physical and environmental improvements were made

In all cases where this analysis could be done, it was determined that the physical and environmental improvements, as far as possible, met the main needs of the communities.

If there were enough resources to carry out all the steps planned

In the cases of Minas Gerais, Rio de Janeiro, and Bahia, there were situations where the resources were inadequate, most often causing delays in the urbanisation process. In the case of São Paulo, despite the fact that there had been delays in the release of funds, resources were considered sufficient for the steps envisaged.

If there were social projects to generate employment and income

In most cases, projects to generate employment and income had been carried out, except for *Bairro Córrego da Ilha* (Sabará, Minas Gerais) and *Sete de Setembro* and *Parque Amélia Santa Margarida* (São Paulo).

In *Parque Capuava* (Santo André, São Paulo), several social projects were also deployed during the intervention.

In Rio de Janeiro, at the time of the works in *Vigário Geral*, there was a co-operative of cleaning services and works. In *Vila Mangueiral*, there were social projects linked to the Secretary of Social Assistance.

In Minas Gerais, due to the large problems related to the generation of income in the slums, programs and actions aimed at expanding the job training were developed, and, consequently, there were increased employment opportunities and an improvement in incomes.

In Salvador, there was an integration of urban interventions with actions that stimulate socio-economic practices, linking the physical interventions to social actions and income generation, such as training courses about the formation of co-operatives (preparation of status, legal proceedings legal, etc.), creation and deployment of a soap factory, a project of sports and recreation for young people, expansion of services and programs of basic health, training of community health agents, support for the creation of nurseries, and improvement of programs and services for reducing unemployment and increasing income.

If there had been some research about users' satisfaction and/or evaluation of the correlation between the project and the interventions implemented

There had been some research in *Vila Senhor dos Passos* (Minas Gerais), in the cases from Sao Paulo and from Rio de Janeiro. The cases of Bahia did not have such research.

If there were social work involvement with the community after the completion of the projects

In all cases except for *Bairro Córrego da Ilha* (Sabará, Minas Gerais), there had been social work involvement with the community after the completion of the projects. This work was linked to social projects to generate employment and income, health education projects, environmental projects, and leisure and health activities.

In *Parque Capuava* (Santo André, São Paulo), more social work activities are scheduled for after the completion of the project. Plans have already been developed for leisure, sport, health, and environmental education programs, with visits to parks and water treatment plants. In addition, a council of representatives was established for each sector (3 persons for each of the 6 sectors), including officials from various community programs developed by the Municipality.

In the cases from Rio de Janeiro, several social projects were implemented, which were linked to education, professional training, Information Technology, and other areas.

At *Vila Nossa senhora do Rosário* (Minas Gerais), the urbanisation process was ongoing and was in a less advanced stage than at *Vila Senhor dos Passos*, where the social project was planned and was being implemented. Removals at any stage of deployment were preceded by social actions.

In Bahia, social activities continued during the post-occupation and included community agents, co-operatives, and schools.

Mechanisms of control to ensure the sustainability of the goals and achievements

In *Parque Capuava* (Santo André, São Paulo), the sustainability of the goals and achievements was to be ensured by community monitoring through the Programme of Community Health Agents.

In Rio de Janeiro, Urban and Social Guidance Agencies (POUSO) were installed in the communities in the final stage of work, integrated with teams of architects, engineers, social workers, and community agents who act by educating residents on the importance of preserving public spaces and the equipment deployed. The teams represent the presence of the Municipality in these localities. Moreover, the POUSO developed a town planning legislation for these communities. In order to ensure the presence of public power in the communities, the POUSO has the function of co-ordinating with agencies that are responsible for the collection of garbage, the installation and maintenance of power and water supply networks, and other public services. The POUSO team guided new construction to ensure that it is made in public areas or places of risk, thereby maintaining the alignment of streets. The work of the technicians also aims to prevent the growth of slums.

In Minas Gerais, in *Vila Senhor dos Passos*, actions to minimise risks had been taken, such as maintaining and intensifying the permanent training of leaders, seeking to integrate new representatives into the group of reference, stimulating the expansion of areas of participation and commitment with the community from its residents, and strengthening the organisational foundations of the community. Moreover, there were actions related to health education and training of community agents to stimulate other people to participate of sustainable issues. Regarding the financial sustainability of families, other actions were also taken, usually linked to programs to generate income.

In Bahia, the activities of the Social Sector of CONDER worked during the post-occupation with community agents, aiming to strengthen the creation of employment and income of the populations in these communities.

3. Conclusions

From the integrated analysis of the research, several conclusions can be made. The main findings from the set of experiences of urbanisation in the case studies are highlighted below.

- The interventions that were developed through programs of urbanisation have been more successful than those that did not have guidelines to be followed.
- The main objective of the programs was generally linked to the improvement or construction of major urban structures in slums, with actions related to social development in a sustainable way, with the potential for community and social integration.
- The programs, in general, can be considered systematic, and it is recommended to avoid isolated actions in a dense urban structure, which may result in inadequate consolidation of facilities.
- All the programs studied were applied to several settlements with different physical and social characteristics. Furthermore, the employment of the same program in various situations allowed it to be improved, and the improvements were applied in later applications.
- The documentation process for completion of the program is important in order to direct the actions to be undertaken. It is necessary to establish responsibilities and procedures of the elements involved.
- In both Rio de Janeiro and Minas Gerais, municipal bodies responsible for the housing sector were created. The decision to establish a technical and administrative group (SMH and URBEL) that incorporated the previous experience of the administrative staff has proved to be extremely fruitful, and it is a fundamental condition for the success of interventions in any public sector of activity.
- In the planning of the projects, the methodologies used in the programs are quite similar. They were developed in an integrated manner involving various departments, and they manage the integration and urban impact, degree of consolidation, state land, health, and geotechnical risks, among other issues.

- The proposed interventions have an integrated and multisectoral nature, presenting the solution of physical, biotic, and anthropic problems. This type of action is therefore of fundamental importance in order to improve environmental conditions of the area and the quality of life.
- In all programs of urbanisation, socio-economic profiles of the communities were drawn up.
- In all programs of urbanisation, the removal and resettlement of housing units is expected when necessary. There is the possibility of working with the population and facilitating the negotiation process; however, studies are needed to address the feasibility of such resettlement.
- The government is the project manager in all programs of urbanisation, and, in most cases, the projects are carried out by specialised companies contracted through bidding.
- A system of monitoring and evaluating projects was designed for only some of the cases.
- The licensing of projects, which involves the examination and approval of projects, was done in most programs of urbanisation, either by municipal and state institutions, but also by other areas of the government.
- In all cases, the public resources were combined with other donors. In the programs studied, the resources were linked to the IDB (Inter-American Development Bank), its local counterpart, the IBRD (International Bank for Reconstruction and Development), and the Participatory Budget, MAE (Ministry of Foreign Affairs of Italy).
- In almost all programs of urbanisation, the institutions used the outsourcing of services, projects, and works as a basic standard through the bidding process.
- Clearly, the guidelines were continually improved in cases where it is possible to monitor the operations.
- It is important to emphasise the important role played by urban regularisation in the transformation of the citizenship conditions of the population.
- There were problems related to the final approval of houses since most of the buildings in these settlements have been built illegally, with no approval from the local government.

4. References

Abiko, A. K. & Almeida, M. A. P. (2000). *Indicadores de Salubridade Ambiental em Favelas Localizadas em Áreas de Proteção aos Mananciais: o Caso da Favela Jardim Floresta.* Boletim Técnico da Escola Politécnica da USP, São Paulo, 32 p.

Almeida, M. A. P. (1999). *Indicadores de Salubridade Ambiental em Favelas Urbanizadas. O Caso de Favelas em Áreas de Proteção Ambiental.* D.Sc. Thesis, Escola Politécnica da Universidade de São Paulo, Dep. de Engenharia de Construção Civil, 226 p.

Bassul, J. R. (2005). *Estatuto da Cidade: Quem Ganhou? Quem Perdeu?* Senado Federal, Subsecretaria de Edições Técnicas, Brasília

COBRAPE (2001). *Programa de Saneamento Ambiental da Bacia do Guarapiranga.* Companhia Brasileira de Projetos e Empreendimentos, São Paulo

Conde, L. P. & Magalhães, S. (2004). *Favela-Bairro: Uma Outra História da Cidade do Rio de Janeiro: 1993/2000. Uma Ação Urbanizadora para o Rio de Janeiro.* Vivercidades, Rio de Janeiro, 157 p.

Denaldi, R. (2003). *Política de Urbanização de Favelas: Evolução e Impasses.* D.Sc. Thesis, Faculdade de Arquitetura e Urbanismo, Universidade de São Paulo

FINEP (2007). *Projeto Refavela.* Relatório Final, MCT

Lareau, S. (2005). *Método para Estimativa de Custos de Infra-Estrutura em Urbanização de Favelas no Município de São Paulo.* M.Sc. Dissertation, Instituto de Pesquisas Tecnológicas do Estado de São Paulo

Santos, C. N. F. (1977). *Volviendo a Pensar em Favelas a Causa de Las Periferias.* Nueva Sociedad, Caracas

Turner, J. F. C. (1972). Una Nueva Visión del Déficit de Vivienda. In: Lewis, D., *El Crescimiento de las Ciudades,* Gustavo Gilli, Madrid

URBEL (2002). *Plano Global Específico: Vila Nossa Senhora do Rosário.* Compania Urbanizadora de Belo Horizonte, Belo Horizonte

12

The Use of Simulation in Urban Modelling

Rosane Martins Alves, Carl Horst Albrecht
Federal University of Rio de Janeiro (UFRJ) - Urban Engineering Program
rosane.alves@poli.ufrj.br, carl@poli.ufrj.br
Brazil

1. Introduction

The impact caused by the fast urban growth and by the occupation of ambient protection areas, demand efficient problem evaluation tools that may be capable to give support in the process of cities territorial, social and economics planning in a short and average stated period. One of the most serious problems of several cities is the disordered urban expansion, which becomes worse because of the lack of planning and specific strategies for this control. Urban systems are becoming ever larger and increasingly complex as urban economies, social and political structures and norms, and transportation and other infrastructure systems and technologies involved.

We intent to provide a reasonable understanding of the context and objectives for urban simulation modelling, the limitations and challenges of urban simulation models, the design choices involved in developing operational models, and how such models are applied.

The process of urban growth, however, is complex and difficult to model, due to the great number of operating actors in the city and the integrated landscape in different scales. The cities size changes all the time and all of possible rules that could be established are extremely complex (Allen, 1997, and Wu, 2002).

To simulate operations and reactions of these real world processes, models of urban environment and the involved actors, are used to assist in exploration of the hypotheses, analysing the ambient processes and giving some answers about urban changes. To deal with these changes it is necessary to know the processes that caused them and identify the conditions. It is important to know how the changes of a city can occur. Cities are non linear complex systems, and their characteristics are difficult to be modelled by conventional methods (static and linear). New strategies have been developed for this class of problems.

New methods of system modelling must be used in modelling and simulating urban phenomenon. Among others, cellular automata and multi-agent systems are being used successfully in cities simulation. CA is simple and is a well established method.

Since the 1980's, the development of discrete choice modelling and the emergence of cellular automata and multi-agent simulation techniques have created a proliferation of modelling approaches. We discuss each of these approaches below, and the supporting role of Geographic Information Systems and the integration of several of these approaches in the design of urban simulation. In this work, a CA urban simulator model will be presented.

2. Modelling Urban System

Some isolated urban process and phenomena can be modelled and simulated by well known methods: Geographic Information System (GIS) for spatial data, virtual reality or computer graphics for architectural issues, finite elements methods for structural and mass flux problems. But a tool for an overall description of the relationship among these process and problems is not completely developed, and is a great research field.
Besides of the complexity of these process and the difficulties in understand, modelling and simulating the interaction between them, urban problems are dependent of the scale. People, for example, can be modelled as a person, with wishes, behaviour all individual thoughts or can be modelled as a mass, reacting to external forces.
Some methods have been developed to carry out the task of simulate urban process and their interactions, such as: Cellular Automata (CA), Multi-agent Systems (MAS) and Artificial Life (AL). Cellular Automata are simple, flexible and has been shown good results (Benenson & Torrens, 2005).

3. Simulators, Agents and Models

A simulator is chosen, following the premise of that it is possible to carry out complex tasks (simulation of urban growth, etc.), through the work carried out for a great number of agents, acting in set, but, separately they are considered very simple.
Agents are computational systems that inhabit some environment of complex dynamics. They feel and act independently in this environment, and carry through a series of objectives or tasks for which they had been assigned. Agents operate without human beings or others direct intervention, and have some control on its action and internal states, perceiving the environment and answer the changes, taking initiatives to reach an objective.
Models of physical iterations are used and are the base of agents modelling, allowing iterations as the attraction and repulsion, which are used in the simulator.
The problem is treated as being a set of agent's interaction. Each one has its own objectives (behaviours) and act on layers, representing the city characteristics in study (eg. railroads, highways, rivers). The exit map is generated as a resultant of the simulation. Figure 1 illustrates the simulator structure.

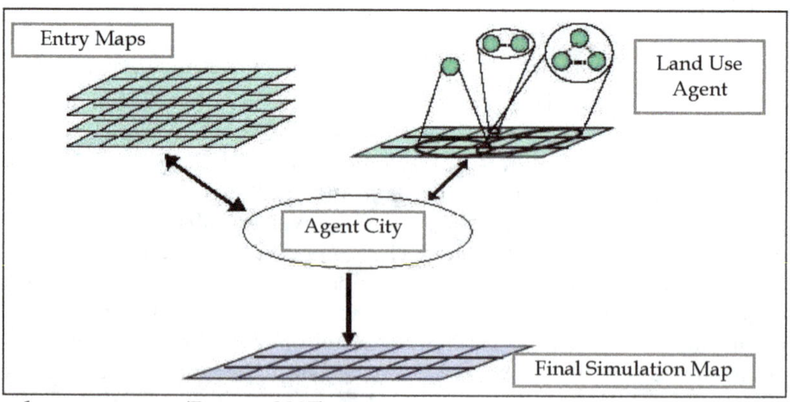

Fig. 1. Simulator structure (Bastos, 2007)

In relation to the model, for understanding the changes that occur in a city, it is necessary not only observe an isolated moment, but also a process of transformation throughout one determined period of time (Portugali, 2000).
Artificial life, multi-agent systems and cellular automata are good examples of simulators.

4. Artificial Life

The field of artificial life (or A-life) is concerned with human-made systems that exhibit life-like characteristics and behaviours. As well as exploring the possibility of creating artificial life, the field is focused on understanding natural life by attempting to abstract the fundamental dynamical principles that underpin biological phenomena - in short, this is a search for the rules that make life possible. A-life has close parallels with complexity studies. Life and living organisms represent some of the best examples of complex adaptive systems. Very simple genetic rules applied across many cells spawn advanced biological structures and organisms. The quest for artificial life is being pursued by replicating the complex dynamics of living systems in other physical media that make them accessible to new kinds of experimentation and testing (Sipper, 1997). In this sense, it is hoped that some of the principles that govern real life can be uncovered in the process. An obvious candidate for such a laboratory is the computer.

5. Multi-Agent Systems (MAS)

Agents are also automata and, thus, incorporate all of the features of basic automata that have just been discussed. However, there are some important distinctions between CA and MAS, particularly when agents are specified with mobility, which is the common interpretation in geographic models.
According to Batty (1976), CA is capable of diffusing state information to neighbouring. However, the individual cells of CA remain fixed in their simulated spaces, they cannot change location. Mobile agents transmit information by themselves, moving to another location, which can be at any distance from an agent's current position. Agents' spatial behaviour can manifest more complex forms than simple relocation. For example, landlord agents might perform spatially mediated sale and purchasing of real estate; the spatial behaviour of agents designed to represent car drivers could include the choice of links and turning opportunities at junctions.
Generally, agent automata employed in social science research (Epstein, 1999, and Kohler, 2002) are used to represent individual decision-makers (or, sometimes, groups of decision-makers). Consequently, the states that are attributed to social science agents are usually designed to represent socio-economic characteristics, and agent transition rules commonly correspond to human-like behaviours. For the most part, however, work in agent-based simulation in the social sciences outside geography is non-spatial in nature, as are the tools that are used. Many of the decisions and behaviours of geographic agents are spatial in nature, and this distinguishes agent tools for geographic applications. MAS, allow a non linear dynamic development (Portugali, 2000).
Multi-agent simulation (MAS) models are related to CA in that both draw on complex systems theory, but, MAS differ from CA emphasis on emergent system behaviour arising from interactions between agents. Research and testing of MAS models accelerated rapidly

after the SWARM software environment (Swarm, 2000) was developed for implementing models of this type. The MAS approach is gaining substantial research interest across the social sciences, since it opens new avenues to analyse social behaviour from an interactive perspective. In economics, the adoption of MAS has come to be known as Agent-Based Computational Economics (Tesfatsion, 2000).

6. Cellular Automata (CA)

Cellular Automata simulation is a useful tool in addressing long-term environmental issues. For a long time, urban modellers have been pursuing a "scientific" way that is intuitively equivalent to "precise", of description and prediction. However, similar to engineering context, as the complex of the system increased, the useful information provided by traditional mathematical models is declining rapidly. This is particularly true in the case of long-term environmental issues, where decision making is characterised by the combination of complexity and uncertainty (Benenson & Torrens, 2005).

In urban grow process, usually, the decision of developing a particular site is made individually by development projects, without co-ordination between projects. Land uses sometimes leads to contradictory policies related to development and preservation. All these features suggest that a simulation approach under a self-organising paradigm is much more appropriate. Through simulating different "rules" the model can generate alternative urban scenarios that may reveal the risk associated with certain development policies, thus allowing necessary precautions to be taken against disastrous consequences.

In recent years there are many researchers have shown interest to analysis and design techniques for complex systems. Cellular automata (CA) are one of the effective methods. Although Cellular Automata (CA) was proposed firstly by Von Neumann and Ulam, from theoretical point of view, in the late 1940's, John Horton Conway's "game of life" ensures the new idea of its application in the computing field (Toffoli & Margolus, 1987). CA are henceforth considered as powerful modelling approach for complex systems in which global behaviour arises from the collective effect of many locally interacting simple components. Subsequently several tools based on CA are proposed to provide meaningful results for real world applications.

Perceiving the city as an open, complex, far from equilibrium and thus self-organised system, we can understand how the global pattern of the city is constructed from uncoordinated local decision-making processes. Cellular Automata (CA) provide a way to simulate such a self organised process. Through development decisions being made on the basis of individual sites, a complex urban pattern can be emerging. The decision of developing a particular site is affected by the pattern in the immediate past. In other words, development is proceeding through discreet interactions during which urban space is constantly evolved (Wolfram, 1984, and Wolfram, 1986).

Interaction among developments is confined within a limit of proximity which is measured by a neighbourhood space. No pre-knowledge of a global pattern exists to guide the direction towards the city is evolved. The transition rule is simple in the sense that it is applied simultaneously to all development sites. Moreover, any rules modifications are being applied instantly to all sites (Wolfram, 1984, and Wolfram, 1986).

Simulation is particularly useful when the issue under question becomes less "predictable" due to its complexity, uncertainty, and non linear iterative natures. The value feature of CA simulation is not its "predictive" power, because the property of a self-organising system is that it is largely unpredictable and uncontrollable (Wolfram, 1986), (Toffoli & Margolus, 1987). Given identical initial conditions, each CA simulation run is unique and never fully repeats itself (Portugali & Benenson, 1995). Although the simulation is unable either to replicate or to predict exact development patterns, it can reveal some qualitative features inherent in the evolution of the system, e.g. the overall rate of land lost. This is largely because the final state of CA is controlled by a set of transition rules. Through linking the rules with their consequences, the model can provide "artificial planning experience" (Portugali & Benenson, 1995) to suggest alternative scenarios of urban growth.

Summarily, the simulation involves the following aspects (Wolfram, 1986): selection of appropriate states in the neighbourhood which are relevant to a particular transition; planning of criteria based on the concerned states to reflect stimulus or constraint to the particular transition; inference of the truth of the particular transition from criteria according to the specified decision-making process; and comparison of all possible transitions associated with a cell and to decide the transition of the cell.

6.1 The Evolution of CA

In the beginning, the early development of the CA framework took place in the 1950s and 1960s and is generally associated with famous names and great discoveries of the twentieth century. (Benenson & Torrens, 2005). Cellular Automata in their classic sense were invented by Ulam and Von Neumann in the mid-1940s. They were interested in exploring whether the self-reproducing features of biological systems could be reduced to purely mathematical formulations (Sipper, 1997). At that time, the two worked at Los Alamos Laboratories on the atomic and, later, hydrogen bombs and Stanislaw Ulam, together with Edward Teller, signed the patent application for the latter. Mathematical folklore attributes the CA idea to Ulam, who had exceptional mathematical imagination and avoidance of writing. Although there are doubts about the origins of the idea: "one can say that the "cellular" comes from Ulam and the "automata" comes from Von Neumann" (Rucker, 1999). By 1943, Ulam suggested the idea of cellular space, where each cell is an independent automaton, interacting with adjacent cells, and shared the idea with Von Neumann. The common view, now, is that Ulam's idea was also secondary one, and was based on paper by Alam Turing, (1936), where he demonstrated that a simple automation, later termed a "Turing machine", can simulate any discrete recursive function. Regardless of the origins, CA came into being amid a soup of very talented intellects.

Having being responsible for researching some of the most critical defence projects of World War II, Ulam and Von Neumann did not care too much about publishing their theoretical thoughts. Most of the papers by Von Neumann on CA were completed and published after his death, in the 1960's (Taub, 1961), (Burks, 1966). The first paper by Von Neumann, "The General and Logical Theory of Automata", introducing what are now know as cellular automata, was published in 1951 (Von Neumann, 1951), and discussed the problem of designing a self-reproducing machine.

The developed urban models in the end of 50's until the half of 80's, in a general way, did not operate on a space dimension. The urban space was disaggregated in units (generally zones of origin-destination), but, the result of these models could not be visualised in space. In fact, effective advances in urban models space representation occurred only in the end of 80's, when models of cellular automata had started to be used on a large scale.

Stephen Wolfram, one of the most famous theoreticians defines the Cellular Automata as being a mathematical idealisation of physical systems, in which space and time are discrete, and the attributes assume a set of discrete values too. A cellular automata is a regular uniform grating or matrix field, generally infinite in its extension, with one discrete variable in each locality (cell), evolving in discrete spaces of time. The variable value in one cell is affected by the values in cells neighbouring, found in the previous time step. Each cells variables are brought up to date simultaneously, based on the neighbouring variable values in previous time step and in agreement with a set of pre-defined local rules (Wolfram, 1983).
CA models have applications in most different areas, since in physic until changes in land use and covering, engineering and traffic control, dissemination of epidemics, biology, among others. CA had been, in a implicit way, in the first generation of computational models in 60's with experiments, executed in North Carolina. In 70's, Tobler, influenced by quantitative geography, suggested cellular models for the development of Detroit. Shortly afterwards he started to explore the form through which CA could be applied to geographic systems, resulting in his famous article " Cellular Geography" (Tobler, 1979). Finally, on the end of 80's, CA had widely started to be used for urban questions, impelled for the parallel development of the graphical computation and the theory of the complexity, similar chaos and fractals (Batty et al., 1997).
The 90's had lived deeply successive improvements in CA urban models, which had started to incorporate ambient dimensions, partner-economic and politics, and had finally been successful in small and macro scales space. A example of this last case is presented by White (1998), where the demand for residential area use is esteem through a social subsystem that takes in consideration migratory flows between regions, and where the demand for economic activities (industrial, advertising, services) is obtained by means of region subsystems that evaluate the performance of different economic sectors, supplying, thus, parallel, information on job chances, that again are used to compute the residential demand. This model esteem the demand for different kinds of land use, considering the support ambient capacity of the sites in question (natural subsystem), as well as the imposed restrictions in local level for function's, physical's, institution's and infrastructure's aspects.
Theoretical progress in the vast field of artificial intelligence, such as, neural specialists systems, artificial nets and evolutionary computation, which is anchored in the concept of genetic algorithms, recently had been included in target simulations in CA. As showed by (Almeida et al., 2002), just-incorporated methods in CA models, as tools of adjustments of neural nets (Yeh & Li, 2001) and evolutionary learning (Papini et al., 1998), have shown themselves as the most promising for CA next generation urban models.

6.2 CA Models – Basic Concepts

CA models consist in cells arranged in a regular grid that change state according to specific transition rules. These rules define the new state of the cells as a function of their original state and local neighbourhood (Ramos & Silva, 2002).
CA models have three important characteristics: massive parallelism, cellular interactions localisation and basic components simplicity - cells. A construction of a CA model, destined to simulate a specific problem, like the dynamics of population growth, must obey some rules. Among these, the most important are: net geometry, the size of the neighbourhood, the border initial conditions, the states classes and the transition rules (Ramos & Silva, 2002).

The net geometry consists in its form and dimension. In two dimensions there are three types of regular nets (Viher et al, 1998): triangular (Fig. 2), square (Fig. 3), and hexagonal (Fig. 4). In the majority of the cases the square shaped net is used, due to easiness of representation and visualisation.

Fig. 2. Triangular net

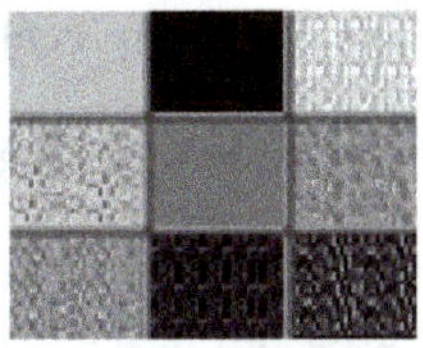

Fig. 3. Square net

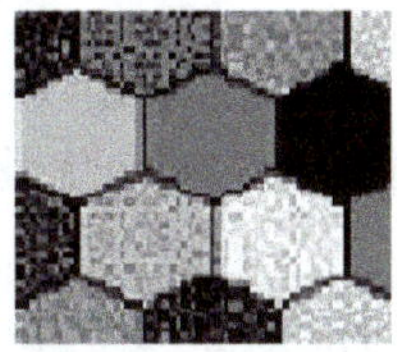

Fig. 4. Hexagonal net

After the definition of the net form, it is chosen the neighbourhood in which the cells can interact. Usually, the models are: Moore neighbourhood, with eight neighbours (Fig. 5), or the Von Neumann neighbourhood (Fig. 6), with four neighbours (Viher et al., 1998).

(a) (b)

Fig. 5. (a) Moore first neighbourhood , (b) Moore second neighbourhood

(a) (b)

Fig. 6. (a) Von Neumann first neighbourhood, (b) Von Neumann second neighbourhood

In the normal CA definition, it is demanded that the net has been defined in all the dimensions, what becomes impossible to simulate an infinite net, truly, in computer. Consequently, it can be prescribed, some border conditions. The initial condition, the classes states and the transition rule, are highly independent aspects (Batty et al., 1997).

The initial condition is the departure scene for the real problem analysis; the cell state classes can represent any characteristic of them, like land use (residential or commercial), population density, among others; and the transition rules, can be determined in other to reflect the way as the real phenomenon happens, and can be interpreted in the simulation, as algorithms. The transition rules specify the behaviour of the cells with time evolution, deciding the future conditions of these cells (Torrens, 2000). Batty et al., (1997) says that, these rules substitute the traditional mathematical functions in the models with procedures based on rules. The author argues, yet, that there are advantages in this methodology: the rules reflect as the real systems operate and allow the reduction of complicated systems in simple ones that have their directed dynamic.

It is important to detach that the GIS, and the graphical technology related to them, supply the necessary platform to increase the complexity of the cellular models, mainly in the study of urban models. Efforts toward deeply understanding about natural phenomena of time-space dimensions have been made. The objective to represent them under the form of dynamic space models by considering future events forecast, consist in promising research areas. Therefore GIS techniques already emphasise the representation of dynamic space phenomena, they are not adequate to foresee future events in changing scenarios.

To represent the relations of interaction, or space tension, between the cells, it is necessary that this structure has been converted into a graph. This is possible because each cell could be considered a vertex of the graph arcs (Granero & Polidori, 2002)(Fig. 7).

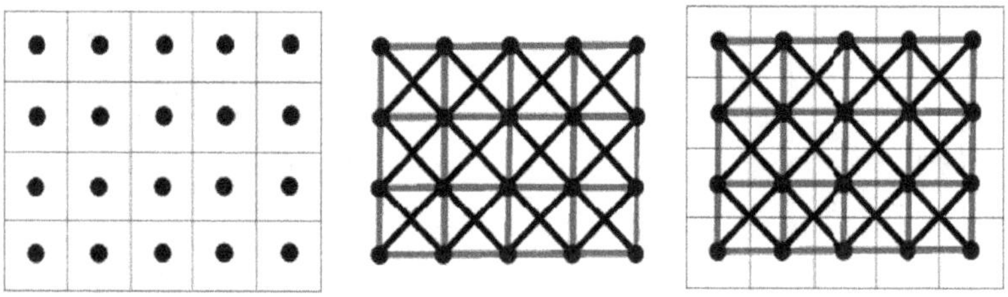

Fig. 7. Cells transformation in graphos (= cells + graphos)

6.3 CA – Some Application Areas in Urban Simulation

Automata systems are the basis of Urban Simulation. Automata-based modelling tools hold many advantages for simulation of urban phenomena in space. The decentralised structure of automata systems, their ability to directly handle individual spatial and non spatial elements, simplicity of formulation, thus, all of these features offer many benefits to model builders (Benenson & Torrens, 2005).

6.3.1 Drainage Network Systems

Although the disadvantages of the urbanisation for the ecosystem and human well-being are known, people are always arriving at urban areas (Geiger, 1993). The growth and development of the cities, many times occurs in a disordered and irregular way, as a consequence of the lack of efficient development plans, supervise and control. This growth leads to a change in land use with greater soil extension watertight.

This increasing of the urban space watertight reflects in cities flooding increase. So, each time more, it is necessary to use tools that make possible urban drainage planners to foresee what could happen, in case of risk's scenarios (as a population increase) becomes a reality.

The numerical simulation appears as a possible tool to be used, allowing the impact evaluation as a consequence of these urbanisation, and from these results to analyse solutions that could minimise the impacts.

The possibility of analysing the impact from different developments scenarios and the combination with the use of tools to control the flooding, become the simulation, in general, a tool widely used in urban drainage managing plans. Normally, the difficulties usually related with the accomplishment of a simulation for urban areas have a relationship with necessary information (drainage network systems, impervious rates, observed runoff, etc.) and with the appropriate choice of the simulation model, and this is conditioned by the available information. However, the ideal would be the possibility of a detailed representation of the urban space, using a model compatible with this proposes. For example, the use of a hydrological model, called Schaake (Schaake, 1971) was presented for the detailed representation of surfaces in urban areas. The concept of source control using on-site detention was used during the simulations. The versatility of this model showed the possibilities for drainage planning in urban areas, mainly those that are in developing.

6.3.2 Application of Space Dynamic Models in the Dynamics of Land Use Change

An increasing number of models for predicting land use change in rapidly urbanising regions are being proposed and built using ideas from cellular automata (CA). Calibrating such models to real situations is highly problematic and to date, serious attention has been focused on the estimation problem. These modelling experiments synthesise various information about spatial infrastructure as the driver of urban land use change. This indicates the relevance of the approach for generating forecasts of growth for Brazilian cities in particular and for world-wide cities in general.

The results obtained with land use change simulations by using CA, have the possibility to be clearly understood by politics, planners and decision-makers, in particular, as well as, the public in general. The dynamics of urban land use models that show to be useful in the identification of the main vectors of urban growth and its general tendencies of land use. In this way, it allows that the local authority power, can command and give a direction to urban growth, as the capacity of ambient support and the infra and superstructure availability at the present and on future (Almeida et al., 2003).

The prognostics of urban expansion supplied by these models are also useful to help local managers, in the establishment of goals for social investments in infrastructure and equipment, as for example the prolongation of ways, expansion in the water and sewer net, creation of new bus lines, construction of day-care centers, schools, hospitals, etc.

Decision-makers of private side can equally benefit from these data modelling, a time that transport companies, fixes and cellular telephony, handle TV, Internet suppliers and others, will have subsidies to define priorities and what could be the intensity to invest. The organised civil society, either through not governmental organisations or quarter inhabitants associations, could use the prognostics, in a way of legitimating claim. In this way, arguments will be based on real expansion trends in short and average stated period.

Batty (1976), displays the key-ideas in relation to the applications and proposals of the urban modelling when affirming that: "...There are many reasons for the development of such models: their object in assisting scientists to understand the urban phenomena, through the analysis and experimentation, represents a traditional objective of science; however, the urban modelling has the same importance when helping planners, politicians and the community in general to foresee, to prescribe and to invent the urban future".

6.3.3 Cellular Automata Models of Road Traffic

Traffic cellular automata (TCA) models, are a class of computationally efficient microscopic traffic flow models. TCA models arise from the physics discipline of statistical mechanics,

having the goal of reproducing the correct macroscopic behaviour based on a minimal description of microscopic interactions.
The performance in measurements on a TCA model's cells lattice, is represented in mathematical notations and these quantities are converted into real-world units and vice versa. There is an extensive account of the behaviour aspects of several TCA models encountered in literature. Already, several reviews of TCA models exist, but none of them consider all the models exclusively from the behavioural point of view.
Some TCA models are used to describe city traffic as a two-dimensional grid of cells, or as a road network with explicitly modelled intersections. Cellular Automata have the advantage of modelling the traffic flow on the microscopic scale of individual vehicles and allow the study of large systems due to a simple type of dynamics (Maerivoet & Moor, 2005).

6.4 Advantages and Potentialities of the Urban Models Based on Cellular Automata

CA models had become popular, because they are easy to handle and has an operational simplicity by generating dynamics that can reproduce traditional processes of changes, beyond containing complexity enough to simulate unexpected and surprising changes, as the observed ones in emergent phenomena. These phenomena are flexible. They supply a structure not overloaded with theoretical assumptions, and that it is applicable to a represented space as a regular grid. These models can be articulated with matrix data, normally used in GIS (Geographic Information Systems).
Although dynamic models have been criticised, due to limitations on a way of capturing the integral inherent complexities to the reality (Briassoulis, 2000), it can be argued in favour of their existence and continuity, because they offer an incomparable way of abstracting standards, dynamic order and trends lines of processes direction of the real world.
As displayed by Batty (1976), "... standard and order exist, in fact, and it is relatively easy to identify them... in urban and regional systems. If a person agrees or not with the description statistics of these standards, it is a question of opinion, and lately of faith in the fundamental ideas."
In the truth, urban models must be conceived, manipulated, applied and interpreted, of a wise and critical form, in a way that the planners and decision-makers, the private and politics sphere, can extract the best of its results and sensibly recognise its limits. These ideas are well synthesised by (Batty, 1976), when affirming that: " a more liberal perspective of the state of the art, of all involved ones, is necessary in the urban modelling, promoting the vision of that models, assisting the imagination inside of a bigger project process and on the solution of the problems and in decision making, in the society as a whole".

7. Conclusions

Urban model is a young field of development, although the beginning studies come from the 50´s. The problem is so vast and complex and we are in the beginning of understanding the complexity of the processes and the interactions between the actors, in modelling urban systems. The rapid development of computers, bringing more and more computational power at lower costs, allows that new forms of exploration can be used in world modelling. Nowadays, bigger and more detailed models are possible, allowing the researchers to improve the complete description of the city behaviour.

The Urban simulation system is being further developed to adapt to varying data availability. Different factors influencing agent choices in locations ranging from newer and rapidly growing.
Careful design at each stage of the process is needed to make the model sensitive to the policies of principal concern, to make the data and computational requirements manageable, to make the model usable by staff and other users with appropriate levels of training, and to fit into the operational practices of the relevant organisations.
To be relevant in the policy process, model design should carefully integrate the elements into a design that fits well into a specific institutional and political context, and evolve to adapt to changing conditions. Careful design at each stage of the process is needed to make the model sensitive to the policies of principal concern, to make the data and computational requirements manageable, to make the model usable by staff and other users with appropriate levels of training.
To deal with these new models, new approaches for simulation have been developed, and CA seems to be the one of the most adequate simulator. It is simple, modular, and easy to implement, and permit that the problem could be represented in almost any scale.
Cellular Automata can take full advantages by using Parallel Processing, a new and powerful computation category.

8. References

Allen, P. M. (1997). *Cities and Regions as Self-Organising Systems: Models of Complexity,* Gordon and Breach Science Publishers, Amsterdam

Almeida, C. M.; Monteiro, A. M. V.; Câmara, G.; Soares-Filho, B. S.; Cerqueira, G. C. & Pennachin, C. L. (2002). Modelling urban land use dynamics through Bayesian probabilistic methods in a cellular automaton environment, In: *29th International Symposium on Remote Sensing of the Environment,* Buenos Aires, Argentina

Almeida, C. M.; Monteiro, A. M. V.; Câmara, G.; Soares-Filho, B. S.; Cerqueira, G. C. & Pennachin, C. L. (2003). Stochastic cellular automata modelling of urban land use dynamics: empirical development and estimation. *Computers, Environment and Urban Systems,* Vol. 27, No. 5, pp. 481-509, September 2003, Elsevier, New York

Bastos, A. D. (2007). Simulação de crescimento urbano utilizando uma abordagem baseada em Sistemas Multiagentes Reativos, *M.Sc. Dissertation in Computational Science,* Instituto de Informática, UFRGS, Porto Alegre

Batty, M. (1976). *Urban modelling: algorithms, calibrations, predictions,* Cambridge University Press, 381p, Cambridge

Batty, M.; Couclelis, H. & Eichen, M. (1997). Urban systems as cellular automata, *Environment and Planning B,* Vol. 24, No. 2, p. 159-164, March 1997

Benenson, I. & Torrens, M. (2005). *Geosimulation: Automata-based Modelling of Urban Phenomena,* John Wiley & Sons, LTD, ISBN: 978-0-470-84349-9, England

Briassoulis, H. (2000). Analysis of land use change: theoretical and modelling approaches, In: *The Web Book of Regional Science* (S. Loveridge, Ed.), West Virginia University

Burks, A.W. (1966). *Theory of Self-Reproducing Automata by John von Neumann,* University of Illinois Press

Epstein, J. M. (1999). Agent-based computational models and generative social science, *Complexity,* Vol. 4(5), pp. 41-60

Geiger, W. F. (1993). Concepts for flood control in highly urbanized areas, In: *IAWQ International Conference on Urban Storm Drainage,* Niagara Falls, Canada

Granero, J. C. & Polidori, M. C. (2002). Simulador da dinâmica espacial com representação de um ambiente SIG, *IV Simpósio Brasileiro de Geoinformática,* Caxambu, Brazil

Kohler, T. A. (2002). Putting social sciences together again: an introduction to the volume, *Dynamics in Human and Primate Societies,* Oxford University Press, pp. 1-18

Maerivoet, S. & Moor, B. (2005). Cellular automata models of road traffic, *Physics Reports,* Vol. 419, pp. 1-64

Papini, L.; Rabino, G. A.; Colonna, A.; Di Stefano, V. & Lombardo, S. (1998). *Learning Cellular Automata in a Real World: The Case Study of the Rome Metropolitan Area*

Portugali, J. (2000). *Self-Organization and the city,* Springer, Berlin

Portugali, J. & Benenson, I. (1995). Artificial planning experience by means of a heuristic cell-space model: simulating international migration in the urban process, *Environment and Planning A,* Vol. 27

Ramos, R. A. R. & Silva, A. N. R. (2002). Oportunidades e desafios de técnicas emergentes para o planejamento urbano – O caso dos modelos de Cellular Automata, *Associação de Utilizadores de Innformação Geográfica,* Lisboa

Rucker, R. (1999). *Seek! Selected Nonfiction by Rudy Rucker,* Four Walls Eight Windows

Schaake, J. C. (1971). Modelling Urban Runoff as a Deterministic Process, In: *Treatise Urban Water Systems,* Colorado State University, p. 343-401

Sipper, M. (1997). Evolution of parallel cellular machines*: The Cellular Programming Approach,* Springer, Berlin

Swarm, J. (2000). www.swarm.org., Santa Fe Institute, Santa Fe

Tesfatsion, L. (2000), *Introduction to the Special Issue on Agent-Based Computational Economics,* Department of Economics, Iowa State University, pp. 1-9

Taub, A.H. (1961). John Von Neumann: collected works, *Design of Computers, Theory of Automata and Numerical Analysis,* v.5, Pergamont Press, Oxford

Tobler, W. R. (1979). *Cellular Geography,* In: Gale, S. & Olsson, G. (ed.) Philosophy in Geography, pp. 279-386

Toffoli, T. & Margolus, N. (1987). *Cellular Automata Machines,* MIT Press, Massachusetts

Viher, B.; Dobnikar A. & Zazula, D. (1998) Cellular automata and follicle recognition problem and possibilities of using cellular automata for image recognition purposes, *International Journal of Medical Informatics,* Vol. 49, pp. 231-241

Von Neumann, J. (1951). The general and logical theory of automata, *Cerebral Mechanisms in Behaviour - The Hixon Symposium,* Villey, pp. 1-41, New York

White, R. (1998), Cities and cellular automata, *Discrete Dynamics in Nature and Society 2,* pp. 111-125

Wolfram, S. (1983). Statistical mechanics of cellular automata, *Reviews of Modern Physics,* Vol. 55, pp. 601-644

Wolfram, S. (1984), Cellular automata as models of complexity, *Nature,* No. 311, pp. 419-424

Wolfram, S. (1986). Theory and applications of cellular automata, *World Scientific Publishing,* Singapore

Wu, F. (2002). Complexity and urban simulation: towards a computational laboratory, *Geography Research Forum,* pp. 22-40, England

Yeh, A. G. O. & Li, X. (2001). A constrained CA model for the simulation and planning of sustainable urban forms by using GIS, *Environment and Planning B,* Vol. 28

13

Spatial Analysis for Identifying Concentrations of Urban Damage

Joseph Wartman, Nicholas E. Malasavage
Drexel University Engineering Cities Initiative (DECI)
joseph.wartman@drexel.edu, nem23@drexel.edu
United States of America

1. Introduction

Disasters resulting from earthquakes, hurricanes, fires, floods, and terrorist attacks can result in significant and highly concentrated damage to buildings and infrastructure within urban regions. Following such events, it is common to dispatch investigation teams to catalog and inventory damage locations. In recent years, these data gathering efforts have been aided by developments in high resolution satellite remote sensing technologies (e.g. Matsuoka & Yamazaki, 2005) and by advances in ground-based field data collection (e.g. Deaton & Frost, 2002). Damage inventories are typically presented as maps showing the location and damage state of structures in part or all of an effected region. In some cases information on the post-event condition of major infrastructure systems such as transportation, power, communications, and water networks is also included. Depending on the means used to acquire data, damage inventories may be developed in days (satellite-based data acquisition) or weeks-to-months (ground-based damage surveys) after an event. Once available, these inventories can be used for a range of purposes including guiding emergency rescues (short-term use), identification of neighborhoods requiring post-disaster financial assistance (intermediate-term use), and support of zoning, planning or urban policy studies (long-term use). An important task when analyzing these inventories is to identify and quantify damage concentrations or clusters, as this information is useful for prioritizing post-disaster recovery activities. Additionally, an understanding of damage concentrations can provide insight to the multiscale processes that govern an urban region's performance during an extreme event.

In some cases spatial patterns and clusters can be inferred from damage inventories using simple, qualitative visual assessment techniques. While this may be a satisfactory approach in situations where there is a marked contrast in building performance, its effectiveness is limited when damage contrasts are subtle, and spatial patterns are less obvious. In these instances, more advanced spatial analysis tools such as point pattern analysis can be of benefit.

Point pattern analysis (PPA) techniques are a group of quantitative methods that describe the pattern of point (or *event*) locations and determine if point locations are concentrated (or *clustered*) within a defined region of study. An early and often-cited example of a semi-qualitative application of the PPA concept is physician John Snow's mid-nineteenth century investigation of a cholera outbreak in London (Johnson, 2006). By mapping the locations of drinking water pumps along with the residences of individuals suffering cholera-related illness, Snow was able to link the epidemic to the local water supply. More recently, a rigorous statistical framework for PPA has largely emerged from work within the plant ecology research community. Since the advent of Geographical Information Systems (GIS), PPA has been used with increasing frequency in a range of applications including identification of crime patterns (e.g. Ratcliffe & McCullagh, 1999) and tracking of disease outbreaks (e.g. Lai et al., 2004).

This chapter will review methods from three classes of PPA within the context of an assessment of a high quality building damage inventory. The mathematical formulation of PPA methods have been discussed in detail elsewhere (e.g. Diggle, 2003; Wong & Lee, 2005; Illian et al. 2008) and therefore will not be repeated here. Instead, this chapter will focus on the *application* of PPA techniques and the *interpretation* of results for an urban damage inventory compiled after the 2001 Southern Peru earthquake. Results of the analyses will be compared and discussed along with other pertinent issues. In fitting with the theme of this volume, this chapter is intended to give readers less familiar with spatial analysis a basic framework for understanding key concepts of PPA. More detailed discussions of the techniques discussed in this chapter can be found in Fotheringham et al. (2000), O'Sullivan & Unwin (2003), Fortin & Dale (2005), Mitchell (2005) and Pfeiffer et al. (2008), among other excellent references. Although the chapter is geared toward urban damage inventories, the concepts presented here are appropriate for a wide range of applications in urban engineering and policy (Table 1). Thus it is hoped that this work will inspire more frequent and innovative use of spatial analyses in urban engineering practice and research.

Discipline	Points/Events	Application
Infrastructure Engineering	Underground service repairs	Plan/prioritize maintenance and future upgrades to system
Infrastructure Engineering	Manufacturing centers	Site specific municipal services facilities such as recycling centers
Transportation Engineering	Automobile accidents/pedestrian incidents	Identify roads and intersections requiring safety enhancements
Transportation Engineering	Persons	Siting of transit hubs and connections
Environmental Engineering	Environmental monitoring locations	Identify and track pollution point sources
Civil Engineering	Landslides	Hazard zonation
Public health	Water-borne disease outbreaks	Drinking water quality evaluation
Environmental Science	Urban wildlife sightings	Assess wildlife nesting or migration habits

Table 1. Example of applications of PPA in Urban Engineering and Policy Making

2. Case Study of Damage in San Francisco (Moquegua, Peru) during the 2001 Southern Peru Earthquake

2.1 Overview

The 23 June 2001 moment magnitude (M_w) 8.4 Southern Peru earthquake affected a widespread area that included several important population centers in southern Peru and northern Chile, including Moquegua, the city that will be the focus of this chapter (Figure 1). The earthquake occurred along the active subduction boundary of the Nazca and the South American plates resulting in widespread damage throughout the region. In general, adobe buildings and older structures were most susceptible to damage, though a significant number of modern engineered structures were also impacted by the earthquake. Rodriguez-Marek and Edwards (2003) present a comprehensive overview of the damage caused by the earthquake. Only a limited number of strong motion instruments recorded the main shock, with the largest peak ground acceleration of 0.33 g being measured in the northern Chilean city of Arica. The only ground motion station in Peru, coincidentally located in the city of Moquegua, registered a moderately high peak ground acceleration of 0.30 g.

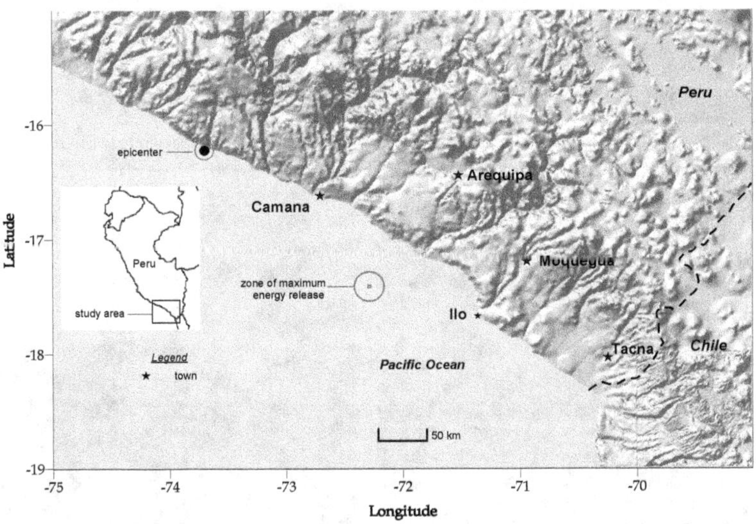

Fig. 1. Regional map showing the location urban centers impacted by the 2001 Southern Peru earthquake

The city of Moquegua (population: 60,000) is situated in an alluvial valley at the base of the Andes Mountains. The city is located approximately 55 km east of the Pacific coast at an elevation of 1400 meters. San Francisco, an approximately 1 km^2 neighborhood located in the southwestern part of Moquegua, was one of the most damaged areas in the city (Figures 2 and 3). In contrast to most of Moquegua, which is relatively flat, San Francisco is distinguished by its variation in topography (Figure 4). San Francisco is situated on a geologic outcrop that includes three ridges rising roughly 100 m above the surrounding portions of the city. This outcrop, which daylights in the upper half of each ridge, consists of stiff conglomerate of the Moquegua geologic formation. This outcrop is also the primary source of alluvium and colluvium that forms a soil mantle that generally thickens with decreasing elevation. Soil thickness ranges from 0 m on the hillside, to approximately 6 m in valley and flatland areas. San Francisco has grown continuously over the past 40 to 50 years to its 2001 population of 12,000. Buildings in the neighborhood are primarily of masonry or similar construction, with a lesser number of older adobe structures. A summary of land use in San Francisco is presented in Table 2.

Fig. 2. Aerial view of Moquegua showing the San Francisco neighborhood outlined in red. A river is visible at north of the neighborhood. (via Google earth, North is vertical)

Fig. 3. Building damage in San Francisco after the 2001 Southern Peru earthquake

The absence of earthquake-induced ground failure (i.e., soil liquefaction and landslides) in San Francisco suggested that the high levels of building damage were a result of strong localized shaking. Several preliminary post-earthquake investigation reports (e.g. Kosaka-Masuno et al., 2001; Kusunoki, 2002; Rodriguez-Marek et al., 2003) hypothesized that the high levels of damage to were due to topographic amplification (Kramer 1996) of ground motion, resulting in localized strong ground shaking. This phenomenon, where topographic features (e.g. hills and ridges) alters and amplifies local ground shaking, has been observed in past earthquakes and is most pronounced near ridge tops. Given the topography of San Francisco, this was a plausible explanation; however, later published data suggested that damage concentrations were located away from ridge tops, indicating that other factors may have governed localized ground motion intensity in the neighborhood. A question then remains: what role, if any, did topography play in the damage distribution in San Francisco? This chapter will consider the topographic amplification question further by conducting a series of analyses to determine if building damage in San Francisco was clustered and if so, to see if the cluster locations coincide with areas of relief as would be expected with topographic amplification.

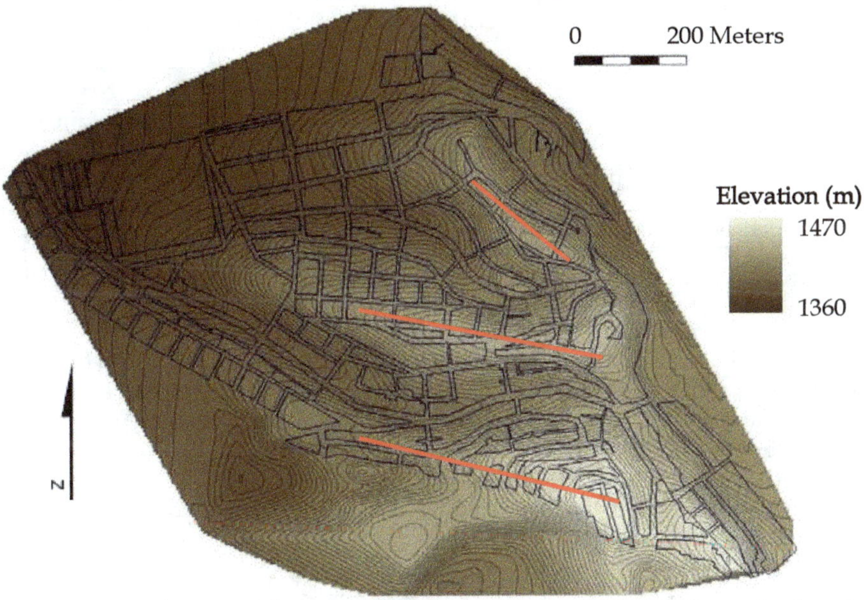

Fig. 4. Street map and topography of San Francisco. The red lines indicate the locations of ridgetops

Land Use	Number of Land Parcels	Percent of Land Parcels
Residential	1611	76.4%
Commercial	89	4.2%
Government	89	4.2%
Vacant	320	15.2%
Total =	2190	100%

Table 2. Summary of land use in San Francisco

2.2 PREDES Damage Inventory

Several earthquake damage inventories for the region have been published, including a high quality, comprehensive account produced by Peru's Center for the Study and Prevention of Disasters (PREDES, 2003). This inventory was developed as part of a larger effort by local engineers, architects and social scientists to assess the effectiveness of short term disaster response in the neighborhood. The PREDES survey was based on individual inspections of close to 1900 buildings whose seismic performance was rated as good, moderate or poor, corresponding to buildings that exhibited no significant damage, significant cracks to loading bearing members, and collapse, respectively. The survey also categorized each building according to its typology (Table 3), as this is known to be an important factor governing the seismic performance of structures. Buildings comprised of masonry and "mixed" construction (i.e., combined masonry and adobe materials) are widely recognized to be of higher construction quality (and seismic resistance) than adobe dwellings.

Building Type	Percentage of Buildings Exhibiting Damage			No. of Buildings (% of Study)
	Low	Moderate	High	
Masonry	81.7%	16.5%	1.8%	968 (52.0%)
Mixed	31.2%	56.6%	12.2%	548 (29.5%)
Adobe	2.6%	21.8%	75.6%	344 (18.5%)

Table 3. Building typology and damage statistics for San Francisco

Figure 5 shows the locations of buildings in San Francisco and their respective typologies. The open areas in the northwest corner of the study area are the locations of two undeveloped land parcels. The study zone was defined to include only the areas where complete data was available for all buildings. This was required because a full data inventory (rather than a sampling) is needed to properly conduct a PPA. Figure 6 shows the post-earthquake damage state of each building in the study area.

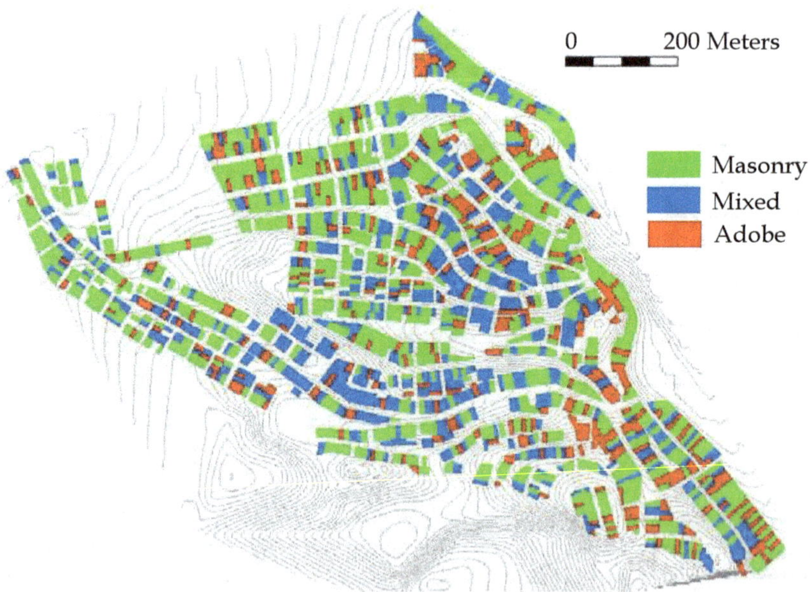

Fig. 5. Building locations and typologies in San Francisco (after PREDES, 2003)

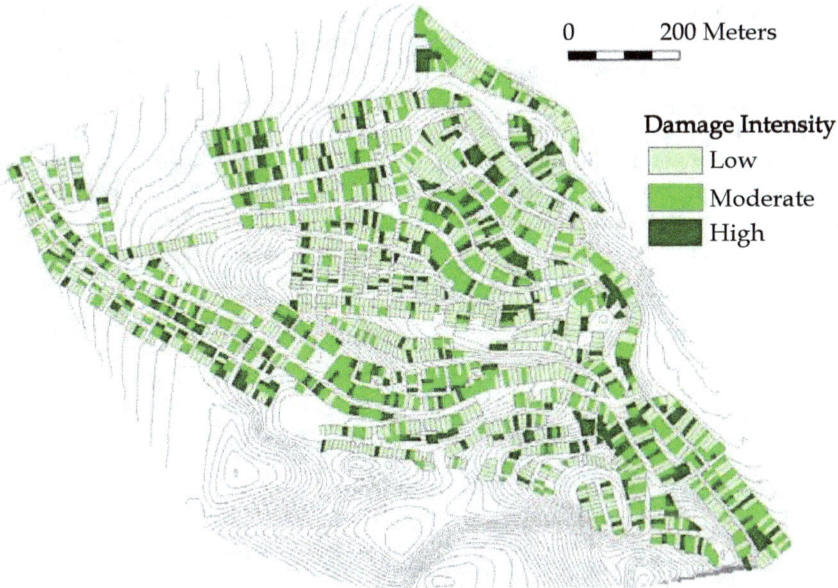

Fig. 6. Post-earthquake damage condition for all structures located in San Francisco (after PREDES, 2003)

Review of Figures 5 and 6 along with the data included in Table 3 indicates the following:

- With some minor exceptions related to a concentration of mixed dwellings located in the south-central portion of San Francisco, the building types appear to be generally well distributed throughout the neighborhood. Most streets are characterized by interspersed masonry, mixed, and adobe dwellings.
- A majority of the buildings in San Francisco are of masonry or mixed construction.
- Overall, a majority of the higher quality (i.e., masonry and mixed structures) buildings in the neighborhood performed well ("low" damage intensity) in the earthquake. In contrast, the seismic performance of nearly all of the adobe structures was poor ("high" damage intensity).
- It is difficult to identify any clear patterns on concentrations of damage from the overall damage survey using visual assessment alone.

The combination of different building types and performance levels make the damage inventory quite complex when considered in aggregate. Although advanced multivariate statistical techniques could be used to analyze the data, PPA provides a more direct approach tailored to the principal objective of the study (i.e., identification of damage clusters). However, this technique requires consistency in the database and thus the PREDES data inventory was modified to ensure uniform building construction quality. Specifically, adobe buildings, which have significantly lower seismic resistance than masonry and mixed structures, were not considered. As these comprised only a small fraction of the total building inventory, this did not significantly reduce the total number of buildings in the inventory. The performance of the remaining 1,513 buildings in the inventory (i.e., masonry and mixed structures) was re-categorized in a binary manner as "no damage/moderate damage" and "collapse". As collapsed buildings (rather than partially damaged) were responsible for most of the injuries and fatalities in the earthquake, the PPA was focused specifically on this damage category. Based on these refinements to the original PREDES database, the point events for the PPA are thus defined as masonry and mixed construction buildings (i.e., higher quality structures) that collapsed in the earthquake. The locations of these damaged buildings, which comprise 5% of the structures in the refined database, are shown in Figure 7. Visual inspection of this figure fails to reveal any marked or otherwise obvious concentrations of collapsed buildings. This inventory of collapsed buildings will be considered in a more quantitative manner in the following sections.

Fig. 7. Collapsed masonry and mixed buildings (shown in red) in San Francisco

3. Point Pattern Analysis (PPA) of Collapsed Building Locations in San Francisco

3.1 Data Requirements for PPA

Point patterns consist of a series of spatially distributed points, or *events*. To constitute a point pattern, a set of events must meet the five criteria highlighted below (O'Sullivan & Unwin, 2003) and discussed in the context of the PREDES damage inventory:

1. The patterns should be mapped on a plane. Owing to the topographic relief in San Francisco, the events occur over a three-dimensional ground surface. Nevertheless, variations in elevation (1360 m to 1470 m) are small relative to the plan dimensions of the neighborhood (approximately 1000 m by 1300 m) and thus the study area can be reasonably approximated as a plane.

2. The study area should be determined objectively, rather than arbitrarily. The boundaries of the San Francisco study area were delineated based on the availability of data; however, this area fully encompasses the major topographic features of the neighborhood and also includes some of the surrounding flatlands. As such, the PPA can be used to determine if clusters are associated with the neighborhood's topographic features.

3. The events must be based on a census of the study area, rather than a sampling. The PREDES inventory was developed based on detailed building-by-building (i.e. census-type) inspections, thus satisfying this criterion.

4. Objects in the study area must directly correspond to events in the pattern. The events were defined so as to directly correspond to masonry and mixed construction buildings that collapsed in the earthquake, thereby satisfying this criterion.

5. Event locations must be proper, rather than representative of a larger object. The damage event locations were taken as the centroid of a building lot; however, as these lots are relatively small (approximately 180 m^2) relative to the size of the study area (approximately 750,000 m^2), this is judged to be a reasonable approach for estimating the event locations.

3.2 PPA techniques

There are three basic approaches for conducting a PPA. The first involves determining the number of events within a given area and therefore are referred to as *density-based* methods. The simplest of this class of methods are quadrat counts, whereby quadrats are drafted over the study area and the number of events in each is counted to determine quadrat densities. A more sophisticated but conceptually similar density-based method involves kernel density functions, whereby events are summed within a series of circular regions centered at a given location within the study area. The second approach for PPA involves the determination of distances between events, and is thus referred to as a *distance-function.* The third approach to PPA considers *spatial associations that vary locally* from the larger global trends across a region of study. As will be discussed below, most of these approaches can be combined with statistical analysis to obtain more rigorous quantitative description of point patterns and clustering.

3.2.1 Quadrat Count Methods

Quadrat counts are perhaps the simplest and easiest of the PPA techniques to understand and use. With this method the intensity of a point pattern (λ) is computed as:

$$\lambda = \frac{n}{\alpha} \tag{1}$$

Where n is the number of events within a quadrat and α is the area of individual equal dimension quadrats imposed over the study area.

Figure 8 shows grids of 50 m, 100 m, and 150 m square quadrats over the study area. Each grid was oriented in a north-south/east-west orientation, with the origin located at the south-west limit of the neighborhood. It should be apparent that grid orientation can affect the resulting quadrat counts, especially when larger grid sizes are used. The color intensity on the figures is proportional to the number of points (events) in each quadrat. Note that the value of the color scale varies between each of the three diagrams.

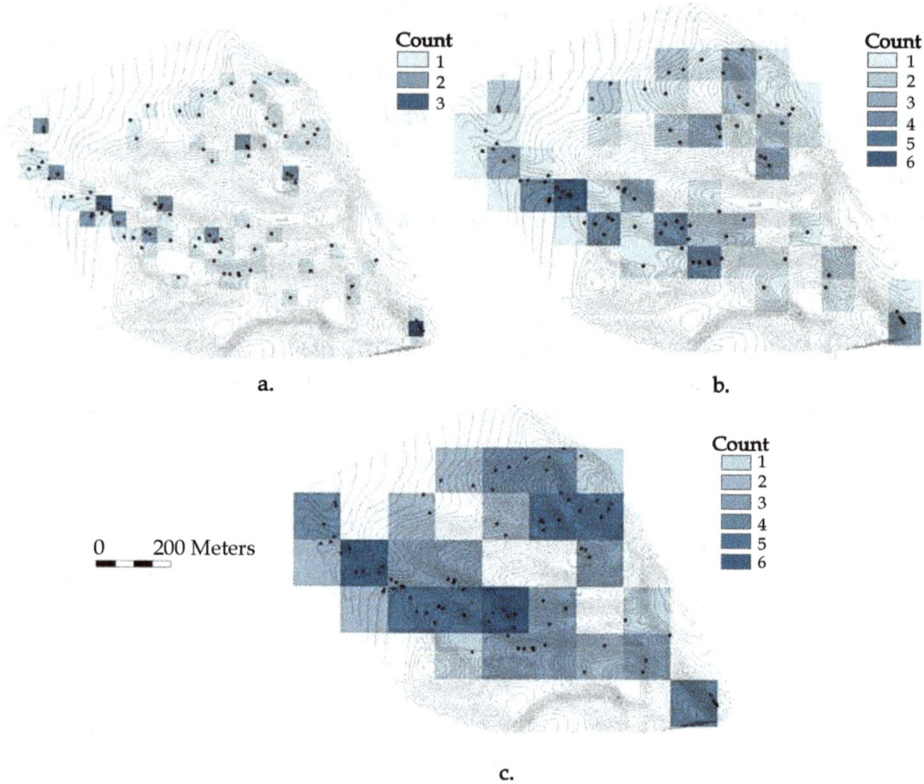

Fig. 8. Results of the quadrat count method for (a) 50 m, (b) 100 m and (c) 150 m square grids

Observe that different quadrat sizes capture different apparent trends in the point patterns. For example, the highest intensity quadrats for closest-spaced grid (50 m) correspond to locations were immediately adjacent or nearly adjacent buildings were damaged, while the largest grid (150 m) captures trends over a significantly larger grouping of dwellings (i.e., street block scale). This observation is not surprising and highlights the scale-dependence of the quadrat count approach. Ideally, the size of the quadrats should reflect the scale of process or mechanism governing the response of the larger spatial system. The question under consideration in this case pertains to the role that relief played in the distribution of collapsed buildings in San Francisco, and thus it is worth considering the scale of the key topographic features in the study area. The local topography is dominated by several ridges and valleys approximately 50 m to 150 m in width, and therefore a reference dimension in this range is appropriate for the analysis. Alternatively, a more general guideline proposed by Greig-Smith (1952) can be used to determine the optimal length (d) of a square quadrat:

$$d = \sqrt{\frac{2A}{n}} \tag{2}$$

Where n is the number of points within the entire study area A.

Applying the Greig-Smith (1952) guideline to the San Francisco damage inventory yields an optimal quadrant dimension of 132 m, which is in the range of the grid spacing considered here. The range of grid sizes considered in Figure 8 is not large, and so it is not surprising that generally similar observations are made from each of the diagrams. Namely, the highest intensity of damage occurs in the (i) west, (ii) south-west, and (iii) north-east portions of the neighborhood. Still, the square quadrats appear awkward when overlaid onto the irregularly-shaped study area making it difficult to visually relate high intensity quadrats to the neighborhood's rounded topographic features (ridges and valleys).

Quadrat counts can be combined with relatively simple statistical analysis to formulate more conclusive statements about the data inventory (O'Sullivan & Unwin, 2003). For example, if one assumes that the events follow a Poisson distribution, the computed and expected intensities can be compared to see if a null hypothesis of a uniform distribution is obtained. Table 4 computes the variance for the 150 m grid quadrat. The mean quadrat count (μ) is 81 events/42 quadrats, or 1.93 events/quadrat. The observed variance can be computed as the sum-of-squared difference (158.21; see Table 4) divided by the number of quadrats (42) to obtain 3.76. This observed variance is a factor of two greater than the mean quadrat count (3.76 > 1), thus indicating a high coefficient of variation among the quadrat counts, which is indicative of clustering. Repeating this exercise for the 100 m (variance/mean = 1.59 > 1) and 50 m (variance/mean = 0.33 < 1) grids, suggesting less pronounced clustering and likely no clustering, respectively. This finding is commensurate with the data presented in Figure 8, which shows greater variability in the intensities (i.e., wider range scale) for the larger grids.

			$\mu =$	1.93
Number of Events (k)	**Number of Quadrats (x)**	k - μ	(k - μ)^2	x (k - μ)^2
0	16	-1.93	3.72	59.51
1	3	-0.93	0.86	2.59
2	6	0.07	0.01	0.03
3	7	1.07	1.15	8.04
4	4	2.07	4.29	17.16
5	4	3.07	9.43	37.73
6	2	4.07	16.58	33.15
Total	42			158.21
			$\chi^2 =$	82.04

Table 4. Quadrat count and variance calculation for 150m grid quadrat

A chi-squared (χ^2) statistical test can also be made to compute the probabilities that the data distribution is clustered. This approach also has the added advantage to associate a level of confidence with the results of the analysis. A chi-squared distribution is expressed as:

$$\chi^2 = \frac{\Sigma(k-\mu_k)^2}{\mu_k} \tag{3}$$

Where μ_k is the mean quadrat count and $\Sigma(k-\mu_k)^2$ is the summation of count difference from the mean quadrat count squared.

Returning to Table 4, the chi-squared value can be compared to standard values for various confidence levels. For the 150 m grid, there are 32 degrees of freedom (i.e. number of quadrats minus 1), and thus the chi-squared values for 99.9% and 99.0% confidence levels, are 62.5 and 53.5, respectively. Comparing these values, it is observed that the actual chi-squared value (79.1) exceeds the both of the critical values, and thus it can be said that the observed quadrat count would be expected to occur in less than 0.1% of the time in random simulations. This high confidence level indicates that the point patterns are very likely to be clustered rather than uniformly distributed. This same statement can also be made for quadrat counts with the 100 m grid ($\chi^2 = 109.8 < 124.3$) and the 50 m grid ($\chi^2 = 313.4 < 368.3$). This finding is commensurate with results of the Poisson-based null hypothesis described earlier.

Before describing the other density method, it is worth noting that there are a number of simple variations on the basic quadrat count theme described here. For example, Wong & Lee (2005) describe the use of hexagon rather than square grids, while O'Sullivan & Unwin (2003) discuss the application of randomly placed sampling quadrats in lieu of the non-overlapping grided census-type described above.

3.2.2 Kernel Density Functions

Kernel density functions comprise a second category of density-based techniques for evaluating point patterns. In the context of spatial analysis, kernel densities are "moving" three-dimensional functions that weight events within a sphere of influence based on their distance to a point where density is being estimated (Gatrell et al., 1996). The basic form of a kernel estimator is:

$$\hat{\lambda}_{\mathrm{p}} = \sum_{i=1}^{n} \frac{1}{\tau^2} k\left(\frac{s - s_i}{\tau}\right) \tag{4}$$

Where $\hat{\lambda}_{\mathrm{p}}$ is the density of the spatial point pattern measured at location s, s_i is the observed ith event, k is the kernel weighting function, and τ is the bandwidth (Borruso, 2008). The simplest of the kernel density functions is the "naive" function, which computes point pattern intensity ($\hat{\lambda}_{\mathrm{p}}$) based on a circle centered at the location where a point density is to be estimated (O'Sullivan & Unwin, 2003). Mathematically, this is expressed as:

$$\hat{\lambda}_{\mathrm{p}} = \frac{no.[S \in C(\mathbf{p}, r)]}{\pi r^2} \tag{5}$$

Where the numerator is the number of events of pattern S within C(**p**,r), a circle of radius r centered at the location of interest **p**. The denominator in this equation (πr^2) is the area of the circle.

The kernel density estimation differs from the quadrat count approach in two key respects: (i) it includes points at but also beyond the location where the intensity is computed, and (ii) kernel density values can be computed at locations throughout a study region to develop contours of intensity values. A benefit of this latter aspect is that it allows discrete point data to be represented as a continuous surface over a study area. These continuous surface values can then be used in direct comparisons with other continuous data fields (e.g. ground surface elevation). The surfaces can also be used to modify or normalize density functions to account for local variations in sample density. For example, in their study of spatial patterns of a disease outbreak, Lai et al. (2004) account for variations in population over the study region by normalizing kernel densities by population density to compute incidences on a per capita basis.

Figure 9 presents kernel density (points/km^2) contours of collapsed buildings for radius values (sample windows) of 50 m, 100 m, and 150 m. Because point densities can be represented as smooth contours, it is easier to visually assess patterns in the study area. The contours of kernel densities based on a 50 m radius (Figure 9) highlight the location where damage occurred, but owing to the relatively small radius, do little to reveal patterns or larger-scale associations between the points. However, kernel density contours based on the 100 m radius begin to show some suggestion of localized clustering along and within a north-west/south-east corridor located in a small valley in the west portion of the study area. This effect is even more pronounced for the data from the 150 m radius kernel density, where the intensity of the points is shown to be quite high along the center portion of the apparent damage corridor.

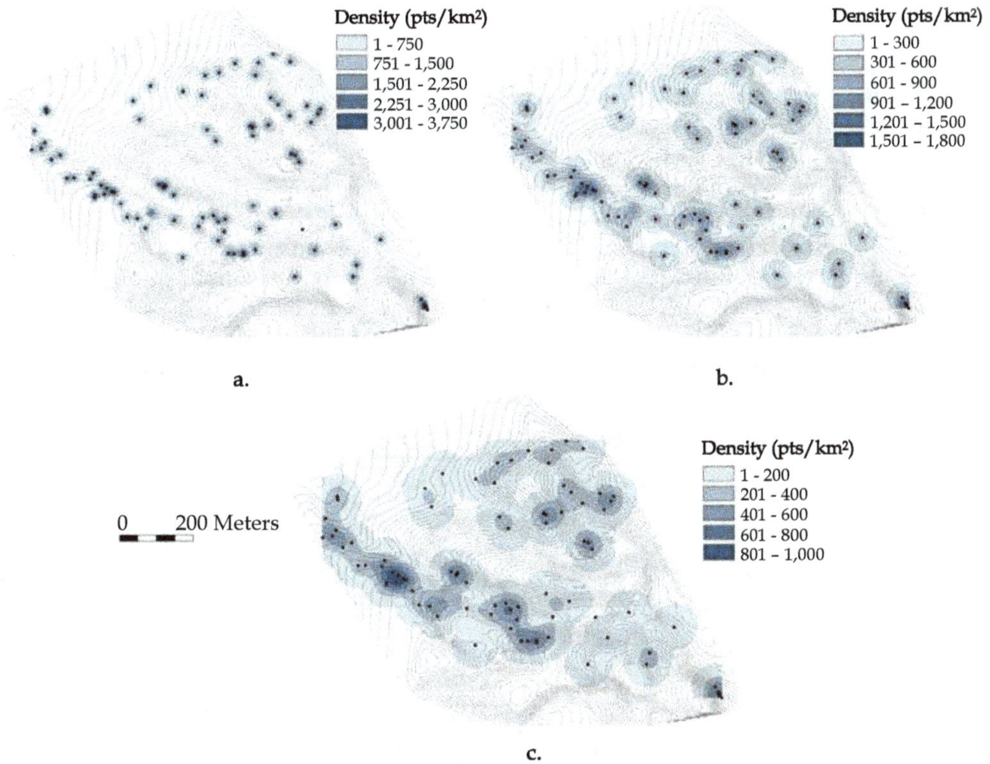

Fig. 9. Results of the kernel density method for (a) 50 m, (b) 100 m and (c) 150 m diameter kernels

Because kernel densities are computed throughout the entire study area, including areas near the periphery where the sampling window may extend into adjacent areas (where point data is not available), artificially low values may be computed in some locations. This phenomenon, referred to as edge effects, has only a limited impact on the San Francisco analysis because very few collapses occur near and beyond the periphery of the study area. Nevertheless, in situations where point clusters exist near the edges of a study area, boundary corrections, such as those discussed by Illain et al. (2008), may be required.

3.3.3 Distance-Based Point Pattern Analyses

Another approach for analyzing point patterns considers the *distance* between points rather than their density. For a single point, this is typically done by measuring the distance to the closest adjacent point to determine the "nearest neighbor" distance. By repeating this for all points, the cumulative frequency distribution (typically referred to as a "G-function") of these nearest neighbor distances can be computed as (O'Sullivan & Unwin, 2003):

$$G(d) = \frac{no.[d_{\min}(p) < d]}{n} \quad (6)$$

Where $no.[d_{\min}(p) < d]$ is the number of events which have a nearest neighbor less than d, and n is the population of the dataset. The value of G for a distance d is, in effect, the fraction of the dataset population having a nearest neighbor less than d.

Plotting the G-function distribution can provide insight to the potential clustering of points, and as with the density-based techniques, the utility of the G-function is enhanced when considered relative to a uniformly distributed or background point pattern. Figure 10 shows the G-function distribution for collapsed buildings located in the study area. The function rises sharply between approximately 10 m and 80 m, indicating that most of the nearest neighbor distances are in this range. At longer distances the distribution continues rising, though at a notably diminished rate before flattening and terminating. An easy qualitative check of the computed G-function distribution may be made by again reviewing Figure 7, which shows that a significant majority of the points have nearest neighbor distances up to about 100 m.

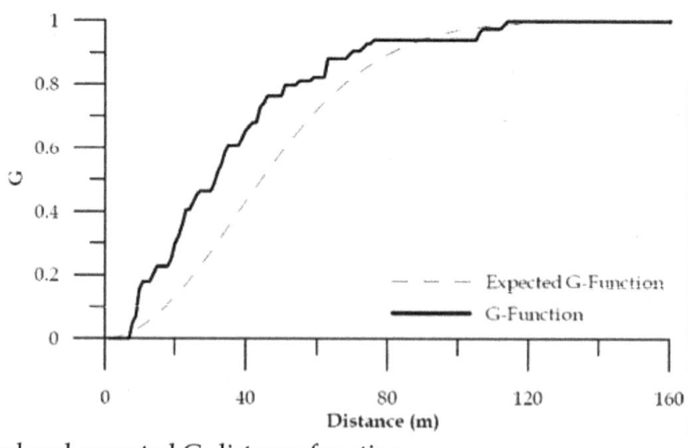

Fig. 10. Observed and expected G-distance function

The dashed line in Figure 10 is the expected G-function distribution for uniformly distributed points. In contrast to the observed data, the expected G-function rises smoothly over the range of distances. Since this function assumes uniform distribution as a null hypothesis, the expected value of G is equal to one when *d* is equal to the study area divided by *n* (i.e. 10 points distributed uniformly in an area 10 m^2 would each have an observed nearest neighbor at 1 m). A key observation from Figure 10 is that the observed G-function distribution is positioned to the left of the expected distribution for nearest neighbor distances up to approximately 65 m. This implies a greater than expected number of nearest neighbor distances over this range, and thus indicates clustering of points.
The K-function (Ripley, 1976) is a popular alternative to the G-function that uses the distance between *all* points (rather than only the nearest neighbor distance) to assess clustering. As such, the K-function effectively considers a distribution of events in proximity to a point being considered. A principal advantage of the K-function is that because it uses all distances it better captures clusters that occur at relatively short distances, which tend to be obscured by G-functions (Ripley, 1976). The K-function is calculated by placing a circle of radius *r* around a point and summing the total number of points located inside the circle. This is repeated for the all points so that the mean count of all events is computed. The mean count value is then divided by the point density for entire study area to obtain K(r):

$$K(r) = \frac{\sum_{i=1}^{n} no.[S \in C(\mathbf{p}, r)]}{n\lambda} \tag{7}$$

Where $\sum_{i=1}^{n} no.[S \in C(\mathbf{p}, r)]$ is the mean of points counted within a circle of radius *r* centered at each point in the dataset, *n* is the number of points in the entire dataset, and λ is the density of points per unit area in the entire dataset.
This calculation is repeated for different radii to obtain a cumulative frequency distribution such as that shown in Figure 11 for collapsed buildings in San Francisco. Interpretation of a K-function is best made in comparison with an expected distribution for uniformly distributed points, or alternatively, against a "background" or control distribution consisting of an entire population (both events and non-events) of points. For San Francisco, the latter entails the location of the entire population of masonry and mixed-construction buildings, regardless of their damage condition, while the former assumes a uniform distribution of points corresponding to collapsed buildings. An advantage of the background over an assumed uniform distribution is that it reflects the actual spatial distribution of vulnerable buildings. Figure 11 includes the K-functions for the collapsed buildings, the control population, and a uniform distribution of buildings. As is typically the case, the arithmetic differences between the observed and expected functions may be small relative to the overall scale of the K-function thereby making it difficult to discern trends from the data (Figure 11). In these instances the data can be presented in other formats to make the trends more apparent. When a control (or background) population is available, a difference function (K_{diff}) can be computed as (Pfeiffer et al. 2008):

$$K_{diff}(r) = K(r)_{damage} - K(r)_{control} \tag{8}$$

Where $K(r)_{damage}$ is the value K(r) for structures with high levels of damage occurred, and $K(r)_{control}$ is the K(r) with respect to all masonary and mixed buildings in the neighborhood.

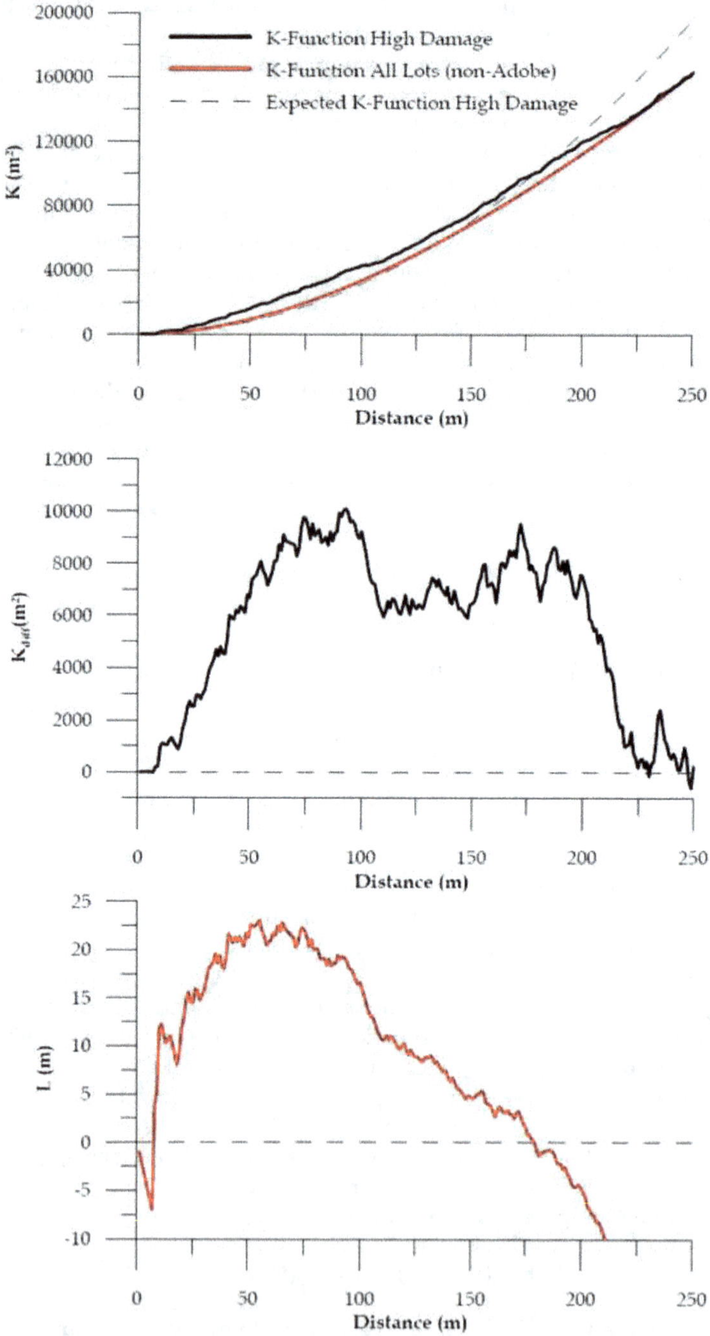

Fig. 11. Observed and expected values for K-, K_{diff}- and L-functions

Another comparison can be made with a uniform distribution of points by computing an L-function (O'Sullivan & Unwin, 2003):

$$L(r) = \sqrt{\frac{K(r)}{\pi}} - r \tag{9}$$

The L-function effectively re-scales the K-function so that observed values plot as zero when they match the expected K-function (i.e., when the observed and uniform K-functions are equal, indicative of no clustering). Figure 11 includes both the L-function and K_{diff} function and for the San Francisco data. The L-function values are greater than the expected value (zero) over the distance range of 10 m to 130 m, with the largest differences occurring between 25 m and 100 m. This is indicative of a high degree of clustering at these distances. At larger distances the values begin to deviate widely in an opposite manner, indicating that the observed points are more widely distributed than expected for a uniform distribution of points. Rather than representing an actual trend in the data, this is a remnant of edge effects, and highlights why it is necessary to consider boundaries when interpreting K-function distributions. In this instance, the deviation occurs because although there are no points beyond the edge of the study zone (thus widely spaced point distributions cannot exist), these areas are nevertheless included in the K- and L-function computations. The K_{diff} function provides additional insight to the clustering of collapsed buildings. Similar to the L-function, the K_{diff} function shows peak values in the range of about 50 to 120 m; however, it also reveals a secondary clustering in the range of about 160 to 200 m. This secondary clustering distance corresponds to the distance between the valleys where much of the damage was in fact located.

3.3.4 Local Indicators of Spatial Association

A more recent approach for identifying and quantifying clusters involves use of local indicators of spatial association, or LISA (Anselin, 1995). The approach is based on the concept of *autocorrelation,* meaning that correlations or relationships exist among spatially distributed data. Most often, data are positively autocorrelated, that is, closely spaced data are more similar then more distantly spaced data. The most popular of the LISA approaches was developed by Getis and Ord (1992), and later modified by Ord and Getis (1995) to what has is commonly known as the Gi* ("G-i-star") statistic. This approach computes the spatial autocorrelation structure over defined sampling windows and compares this to the global autocorrelation structure for the study region. Significant differences between local and global autocorrelation structures imply localized deviations from the global norm, implying localized clustering of events. The Gi* statistic is computed as:

$$G_i^* = \frac{\sum_j w_d^*(i,j)z(j)}{\sum_j z(j)} \tag{10}$$

Where *z(j)* are known attribute values at a known point (i.e. low/moderate damage = 1, collapse = 2) and $w_d^*(i,j)$ is a matrix of spatial weights describing the influence of other known points up to a defined threshold distance (i.e. points within *d* are assigned a weight of 1, while those beyond *d* are assigned a weight of 0).

Similar to other approaches discussed in this chapter, the Gi* statistic is most meaningfully interpreted within a larger statistical framework. In this case, a *Z-score* [Z(Gi*)] can be computed for each point by subtracting the expected Gi* value from the observed value, and dividing this by the square root of the variance for all points in the defined region of study:

$$Z(G_i^*) = \frac{G_i^* - E(G_i^*)}{\sqrt{Var(G_i^*)}} \tag{11}$$

A Z-score of 0 implies no clustering, while increasingly positive or negative values suggest point concentrations. The statistical significances of clustering can be determined by comparing the computed Z-score with a range of values for a given confidence level (Mitchell, 2005). Figure 12 shows the results from this type of analysis for three sampling windows. The filled contours correspond to collapsed building clusters ("hot" spots) associated with statistical confidence levels of 86 % (light red) and 95 % (dark red). The blue contours show the locations of "cold" spots for the same confidence levels.

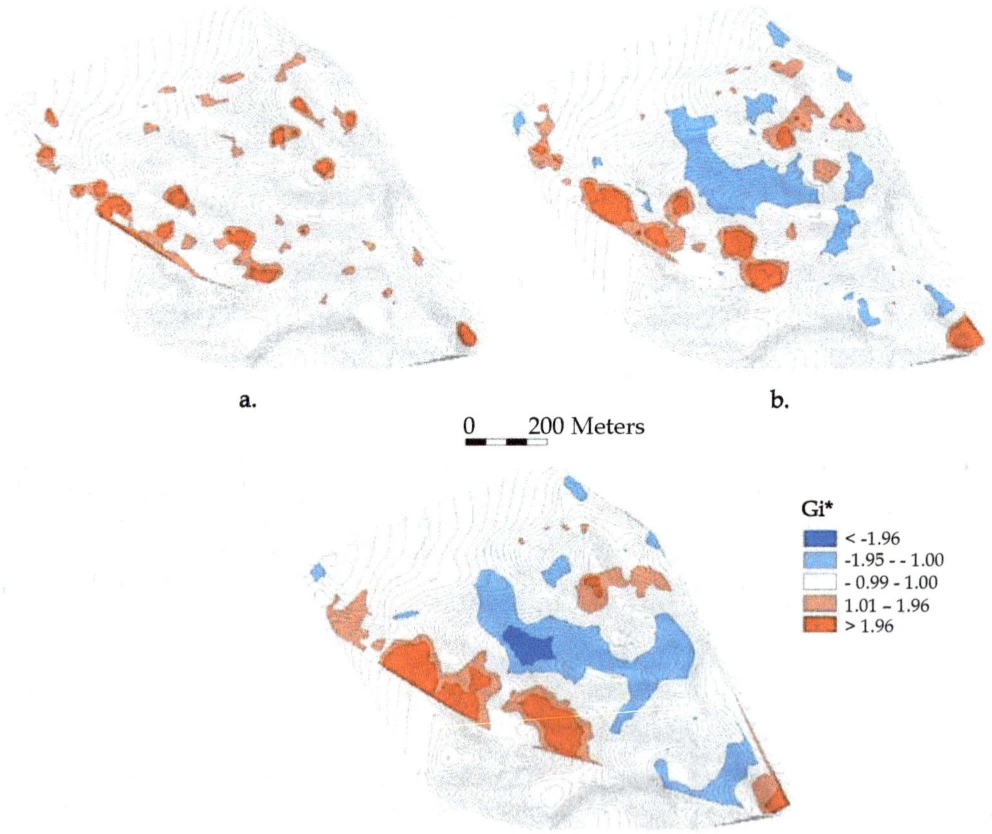

Fig. 12. Gi* value contours for (a) 50 m, (b) 100 m and (c) 150 m diameter sample windows.

Note that the damage clusters shown in Figure 12 are quite similar to those identified from the kernel analysis (Figure 9). A unique and especially useful aspect of a LISA analysis is that extremes at both ends of parameter spectrum can be identified (i.e., "hot" and "cold" spots). For example, included on Figure 12 are blue-toned contours corresponding to local clusters of buildings with a low level of damage (i.e., "cold spots" of low damage caused to mixed and masonry buildings). As with the other PPA approaches discussed in this chapter, LISA-based methods are sensitive to the size of the sample window. As shown in Figure 12, larger sampling windows produce large clusters that become increasingly localized as this dimension is reduced. This localization can become so extreme that some clusters no longer appear; for example, note that the cold spot is not present for the smallest of the sampling windows (Figure 12a).

4. Discussion

Although different in concept, the general categories of PPA discussed in this chapter provide generally similar results for the building collapse analysis in San Francisco (Fig. 13).

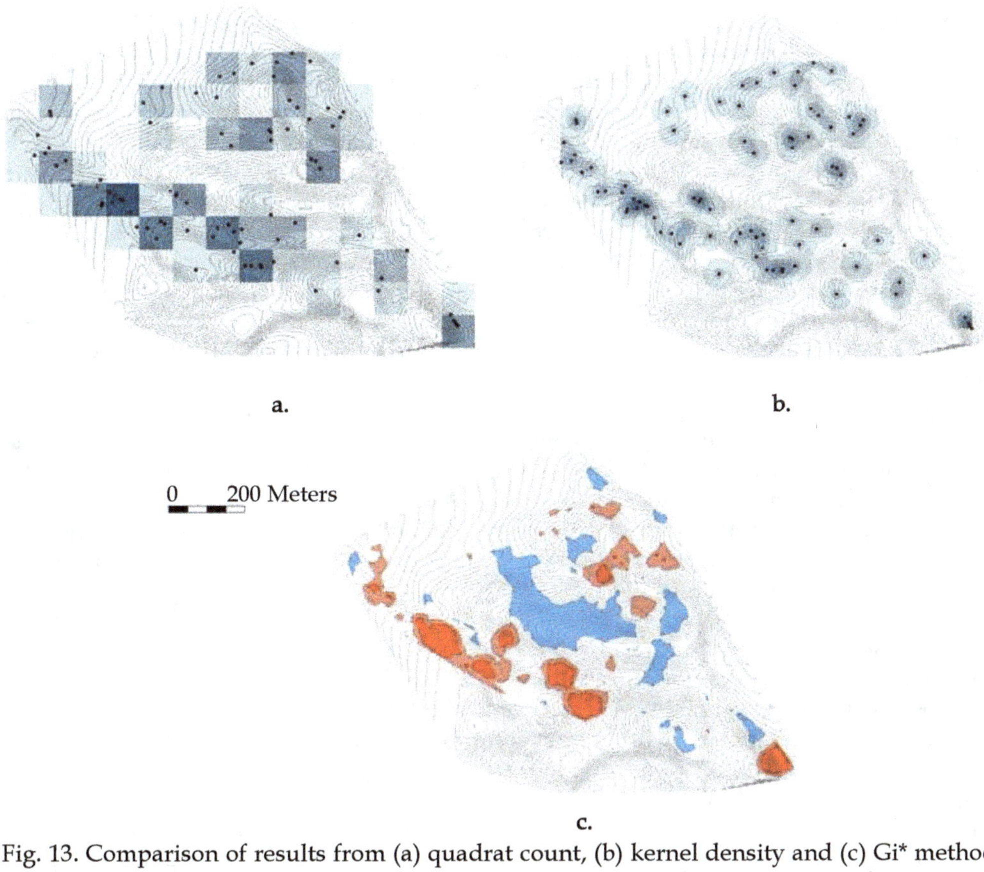

Fig. 13. Comparison of results from (a) quadrat count, (b) kernel density and (c) Gi* method for a 100 m sampling dimension. Results correspond to those of Figures 8, 9 , and 12

These results are:

1. The locations of the collapsed buildings in San Francisco are locally clustered, with the highest degree of clustering occurring for point (i.e., collapsed building) spacings in the range of approximately 25 to 100 m.
2. Referencing Figure 13, collapsed buildings are clustered along a northwest-southeast trending valley. This cluster extends further into an adjacent flatland area in the western portion of the neighborhood. A second and more highly localized cluster of collapsed buildings is found along a hillslope located in the northeast portion of the neighborhood.

These two findings were not fully apparent from the building damage inventory alone. An additional general observation was that for each of the methods, the size of the sampling window used in the analyses can have an effect on the results of the PPA. When small sample windows are specified, highly (perhaps overly) localized clusters are produced, while larger sample windows yield larger and more diffuse clusters.

There are, of course, differences between the approaches, the most important of which pertain to (i) computational effort, (ii) resolution/visual interpretation, and (iii) the nature of the quantitative interpretations that can be made from the analyses. Quadrant counts, the simplest of the approaches, can be determined manually without the need for computer software. Moreover, statistical analysis of quadrant count data is straightforward. In contrast, calculation of the more sophisticated Gi* statistic and related Z-scores is significantly more involved, although modern, commercially available software packages can significantly ease this computational burden.

PPA methods that produce continues surfaces of contours of clusters intensity (i.e., kernel density and Gi* analyses) are relatively easy to interpret visually. In contrast to the quadrant counts, these can be used to identify localized clusters with a high degree of resolution. However, if only the characteristics of the point spacing, rather than the location of clusters are of interest, then the distance based functions (G-, K-, K_{diff} and L-functions) are most useful.

Perhaps one of the more important differences between the PPA methods pertains to the nature of the quantitative interpretations that can be made from the data. The most comprehensive of these interpretations are made based on the Gi* statistic, which provides direct information on both the location of clusters and their statistical significance. While the statistical significance of the quadrant counts and distance based functions can also be determined, these are not directly coupled to high resolution visual presentation of the cluster analysis results.

For the San Francisco case study presented here, the Gi* is perhaps the most powerful of the PPA tools for identifying and quantifying point patterns, though as indicated earlier, the other approaches also provided generally similar results, with less computational effort. Perhaps the most useful application of simpler PPA methods is to provide a check on results from other more computationally intensive approaches. Table 5 summarizes the relative merits of the approaches.

Regardless of the method used for analyses, the selection of the size of the sampling window for a PPA analysis is an important consideration that can have a significant effect on the results. For this reason, when performing a PPA it is prudent to consider a range of scales to better understand the sensitivity of the results to this factor. Ideally, the scale of sampling window should be made relative to a key reference dimension of a process or mechanism governing the response of the spatial system under consideration; however, this is not always known in advance of an investigation. In these instances, criteria developed by Greig-Smith (1952) can provide guidance on an optimal sample window dimension. Edge effects can also have an important influence on the results of a PPA, and measures to account for these may be required for cases where such effects are significant (O'Sullivan & Unwin, 2003).

Finally, note that the collapsed building inventory analyzed in this chapter is static in that it represents damage at a fixed point in time (i.e., damage at the time of the earthquake); it is worth mentioning that PPA can also be used to monitor time-dependant changes in databases having a temporal element.

Category	Method	Benefits	Limitations
Density	Quadrat Count	• Easy to perform analysis using simple hand computations • Can be readably combined with statistical analysis	• Square quadrates can obscure visual association of point intensity with underlying process • Orientation and position of grid can affect results
	Kernel Density	• Provides density surface that can be compared or normalized against other continuous variables	• Does not provide a measure of the statistical significance of results
Distance	G-function	• Quantifies point spacing characteristics	• No information on cluster location
	K-function	• Quantifies point spacing characteristics • Allows for direct comparison of events with background points	• No information on cluster location • Can be highly sensitive to edge effects
LISA	Gi* statistic	• Cluster location directly associated with confidence interval	• Computationally intensive

Table 5. Benefits and limitations unique to PPA methods featured in this chapter

Returning to a central question of this case study: what role, if any, did topography play in the damage distribution in San Francisco? The PPA results clearly indicate that the largest cluster of collapsed buildings is correlated with flatter ground over the study area, which contrasts with preliminary studies that suggested concentrations of highly damaged buildings were localized at ridge tops. This suggests that topographic implication largely governed the performance of buildings. Moreover, damage cold spots identified from the Gi* analysis (Figure 12) generally overlie areas of topographic relief, which further discredits the topographic amplification hypothesis. The clusters in this study are associated with a high degree of statistical confidence, and thus it is unlikely that the observed concentrations of collapsed buildings were a result of natural background variation. These findings by themselves do not fully disqualify the hypothesis that topographic amplification played a role in damage distribution; however, they suggest that at a minimum these effects were less significant then originally thought. It is possible that the damage distribution was more closely related to "site effects" (Kramer, 1996) whereby earthquake ground motion is amplified by surficial soils and sediments such as those that underlie the small valleys and flatlands of San Francisco. Note that the scale of clustering of collapsed buildings (i.e., 25 to 100 m) corresponds closely with the dimensions of the valleys in the study area, lending further credibility to the site effects hypothesis. This brief discussion highlights both a strength and a limitation of PPA: while this can be used to explore spatial data and test hypotheses about larger governing mechanisms, often they cannot by themselves be used to conclusively identify these larger mechanisms and processes. Aside from investigating the topographic amplification hypothesis, the results of the PPA analyses can serve the important practical function of supporting building code zonation in San Francisco. For example, as the neighborhood is rebuilt and repopulated, it would be prudent to use more seismically resistant construction methods for buildings located in the identified collapse cluster zones as these are likely to again experience high levels of shaking in future earthquakes.

5. Conclusions

This chapter introduced three general categories of techniques for identifying and quantifying point patterns in urban engineering applications. These PPA techniques included (i) density-based methods of quadrat counts and kernel density functions, (ii) distance-based G- L-, K_{diff}, and K-functions, and (iii) the local indicator of spatial association (LISA)-based Gi* statistic. Each of these techniques was applied to a comprehensive urban damage inventory developed after the 2001 Southern Peru earthquake. While the application of these techniques in a GIS environment is relative straightforward, careful assessment is needed to meaningful interpret PPA results. This is especially true when considering factors such as the size of the sampling window and the effects of study area boundaries.

Overall, the three procedures all provided generally similar results when applied to the building damage inventory, though differences were noted between each of the methods. The Gi* statistic, which provides both the location of clusters and a statistical measure of confidence, proved to be the most powerful PPA technique for analyzing the San Francisco database. Each technique has its relative merits and ultimately the selection of a PPA methodology should be made with consideration of the larger goals of the investigation. For example, distance-based approaches are best for situations when the characteristics of a point pattern spacing are desired, while kernel density function may be best when point pattern data requires normalized against another data field (e.g. population density). Regardless of the technique chosen for analyses, results should be considered in a statistical framework as this can greatly enhance the quantitative interpretation of the results.

6. References

Ancelin, L. (1995). Local indicators of spatial association - LISA. *Geographical Analysis*, Vol. 27, 93-115

Center for the Study and Prevention of Disasters (PREDES)(2003). Study of the physical environment, social economics and evaluation of the risk in the neighborhood of San Francisco with reference to the District of Moquegua and Mariscal Nieto Province. *Center for the Study and Prevention of Disasters*, 261p.

Deaton, S. L. & Frost, J. D. (2002). *Integrated Digital Earthquake Reconnaissance*, 7th U.S. National Conference on Earthquake Engineering, CD-ROM

Diggle, P. J. (2003). *Statistical Analysis of Spatial Point Patterns*, Hodder Arnold, 0340740701, London, England

Fortin, M. & Dale, M. (2005). *Spatial Analysis : A Guide for Ecologists*, Cambridge University Press, 0521804345, New York, New York

Fotheringham, A. S., Brunsdon, C. & Charlton, M. (2000). *Quantitative Geography : Perspectives on Spatial Data Analysis*, SAGE Publications Ltd., 0761959475, London, England

Gatrell, A. C., Bailey, T. C., Diggle, P. J. & Rowlingson, B. S. (1996). Spatial point pattern analysis in geographical epidemiology. *Transactions of the Institute of British Geographers*, Vol. 21, 256-274

Getis, A. & Ord, J. K. (1992). The analysis of spatial association by use of distance statistics. *Geographical Analysis*, Vol. 24, 189-206

Greig-Smith, P. (1952). The use of random and contiguous quadrats in the study of the structure of plant communities. *Annals of Botany*, Vol. 16, 293-316

Illian, J., Penttinen, A., Stovan, H., & Stoyan, D. (2008). *Statistical Analysis and Modelling of Spatial Point Patterns*, John Wiley & Sons Ltd., 0470014911, Hoboken, New Jersey.

Johnson, S. (2006). *The Ghost Map : The Story of London's Most Terrifying Epidemic and How It Changed Science, Cities, and the Modern World*, Penguin Books Ltd., 1-59448-925-4, London, England

Kosaka-Masuno, M., Gonzales-Zenteno, E., Arias-Barahona, H., Minaya-Lizarraga, A., Faran-Bazan, E., Ticona-Paucara, J. (2001). Seismic Hazard Evaluation in the city of Moquegua. *Convenio UNSA-INDECI. Proyect PER 98/018 PNUDINDECI, San Agustin de Arequipa National University*, Arequipa, Peru

Kramer, S. L. (1996). *Geotechnical Earthquake Engineering*, Prentice Hall, 0133749436, Princeton, New Jersey

Kusunoki, K. (2002). *Report on the damage investigation of the 2001 Atico Earthquake in Peru.* Building Research Institute of Japan. 27 p.

Lai, P. C., Wong, C. M., Hedley, A. J., Lo S. V., Leung, P. Y., Kong, J. & Leung, G. M. (2004). Understanding the spatial clustering of severe acute respiratory syndrome (SARS) in Hong Kong. *Environmental Health Perspectives*, Vol. 112, No. 15, (November) 1550-1556

Matsuoka, M. & Yamazaki, F. (2005). Building damage mapping of the 2003 Bam, Iran earthquake using Evnisat/ASAR intensity imagery, *Earthquake Spectra*, Vol. 21, No. S1, 285-294

Mitchell, A. (2005). *The ESRI Guide to GIS Analysis, Volume 2 : Spatial Measurements & Statistics*, ESRI Press, 158948116X, Redlands, California

O'Sullivan, D. & Unwin, D. J. (2003). *Geographic Information Analysis,* John Wiley & Sons Inc., 0-471-21176-1, Hoboken, New Jersey

Ord, J. K. & Getis, A. (1995). Local spatial autocorrelation statistics : distributional isssues and an application. *Geographical Analysis,* Vol. 27, 286-306

Pfeiffer, D., Robinson, T., Stevenson, M., Stevens, K., Rogers, D. & Clements, A. (2008). *Spatial Analysis in Epidemeology,* Oxford University Press, 9780198509882, New York, New York

Ratcliffe, J. H. & McCullagh, M. J. (1999). Hotbeds of crime and the search for spatial accuracy. *Journal of Geographical Systems,* Vol. 1, No. 4, (December) 385-398.

Ripley, B. D. (1976). The second-order analysis of stationary point processes. *Journal of Applied Probability,* Vol. 13, 255-266

Rodriguez-Marek, A & Edwards, C. (2003). Southen Peru earthquake of 23 June 2001 reconnaissance report, *Earthquake Spectra,* Vol. 19, No. S1, pp. 165

Rodriguez-Marek, A., Williams, J. L., Wartman, J., Repetto, P. C. (2003), Ground motion and site response. *Earthquake Spectra,* Vol. 19, No. S1, 11-34

Wong, D. W. S. & Lee, J. *Statistical Analysis of Geographic Information with ArcView GIS and ArcGIS,* John Wiley & Sons, Inc., 0471468991, Hoboken, New Jersey

Permissions

All chapters in this book were first published in Journal of Allergy, by All chapters in this book were first published in MTUE, by InTech Open; hereby published with permission under the Creative Commons Attribution License or equivalent. Every chapter published in this book has been scrutinized by our experts. Their significance has been extensively debated. The topics covered herein carry significant findings which will fuel the growth of the discipline. They may even be implemented as practical applications or may be referred to as a beginning point for another development.

The contributors of this book come from diverse backgrounds, making this book a truly international effort. This book will bring forth new frontiers with its revolutionizing research information and detailed analysis of the nascent developments around the world.

We would like to thank all the contributing authors for lending their expertise to make the book truly unique. They have played a crucial role in the development of this book. Without their invaluable contributions this book wouldn't have been possible. They have made vital efforts to compile up to date information on the varied aspects of this subject to make this book a valuable addition to the collection of many professionals and students.

This book was conceptualized with the vision of imparting up-to-date information and advanced data in this field. To ensure the same, a matchless editorial board was set up. Every individual on the board went through rigorous rounds of assessment to prove their worth. After which they invested a large part of their time researching and compiling the most relevant data for our readers.

The editorial board has been involved in producing this book since its inception. They have spent rigorous hours researching and exploring the diverse topics which have resulted in the successful publishing of this book. They have passed on their knowledge of decades through this book. To expedite this challenging task, the publisher supported the team at every step. A small team of assistant editors was also appointed to further simplify the editing procedure and attain best results for the readers.

Apart from the editorial board, the designing team has also invested a significant amount of their time in understanding the subject and creating the most relevant covers. They scrutinized every image to scout for the most suitable representation of the subject and create an appropriate cover for the book.

The publishing team has been an ardent support to the editorial, designing and production team. Their endless efforts to recruit the best for this project, has resulted in the accomplishment of this book. They are a veteran in the field of academics and their pool of knowledge is as vast as their experience in printing. Their expertise and guidance has proved useful at every step. Their uncompromising quality standards have made this book an exceptional effort. Their encouragement from time to time has been an inspiration for everyone.

The publisher and the editorial board hope that this book will prove to be a valuable piece of knowledge for researchers, students, practitioners and scholars across the globe.

List of Contributors

Carlos Alberto N. Cosenza, Fernando Rodrigues Lima and César das Neves
Federal University of Rio de Janeiro (UFRJ) Brazil

Giovani Manso Ávila
Federal University of Rio de Janeiro (UFRJ) Brazil

Alvaro Seco and Ana Bastos Silva
University of Coimbra Portugal

Marcelo Gomes Miguez and Luiz Paulo Canedo de Magalhães
Federal University of Rio de Janeiro (UFRJ) Brazil

Alex Abiko
University of São Paulo - Escola Politécnica Brazil

Fernando A. N. Castro Pinto
Federal University of Rio de Janeiro (UFRJ) Brazil

Stefan Siedentop
University of Stuttgart Germany

Armando Carlos de Pina Filho and Renato Dias Calado do Amaral
Federal University of Rio de Janeiro (UFRJ) Brazil

Camilo Michalka Jr.
Federal University of Rio de Janeiro (UFRJ) - Urban Engineering Program Brazil

Adauto Lucio Cardoso and Angela Maria Gabriella Rossi
Federal University of Rio de Janeiro (UFRJ) Brazil

Rosane Martins Alves and Carl Horst Albrecht
Federal University of Rio de Janeiro (UFRJ) - Urban Engineering Program Brazil

Joseph Wartman and Nicholas E. Malasavage
Drexel University Engineering Cities Initiative (DECI) United States of America

Index

www.ingramcontent.com/pod-product-compliance
Lightning Source LLC
LaVergne TN
LVHW051047220926
848485LV00001BA/31
9781639895502